Argo CD in Practice

The GitOps way of managing cloud-native applications

Liviu Costea

Spiros Economakis

BIRMINGHAM—MUMBAI

Argo CD in Practice

Group Product Manager: Rahul Nair
Publishing Product Manager: Preet Ahuja
Senior Editor: Arun Nadar
Content Development Editor: Sujata Tripathi
Technical Editor: Rajat Sharma
Copy Editor: Safis Editing
Project Coordinator: Ajesh Devavaram
Proofreader: Safis Editing
Indexer: Tejal Daruwale Soni
Production Designer: Shankar Kalbhor
Marketing Coordinator: Nimisha Dua

First published: November 2022

Production reference: 1271022

Published by Packt Publishing Ltd.
Livery Place
35 Livery Street
Birmingham
B3 2PB, UK.

ISBN 978-1-80323-332-1

www.packt.com

*To my sons, Tudor and Victor, and my wife, Alina, for giving me the strength
and power to overcome all the challenges.*

– Liviu Costea

*To my sons, Yannis and Vasilis, and my wife, Anastasia, for reminding me every day
that life is a continuous learning process in every aspect.*

– Spiros Economakis

Foreword

In their book, Liviu and Spiros provide an excellent introduction to Argo CD that helps you start using it in a matter of minutes. The book provides great introduction base concepts and the basic vocabulary of both GitOps and Argo CD. Besides teaching about Argo CD itself, the book covers a lot of ecosystem tools that are extremely useful and will prepare you for real-life use cases. Even the basic examples come with YAML snippets, which again will be helpful to solve real-life challenges.

Content gets more advanced and more interesting pretty quickly. You will learn lots of interesting details about advanced Argo CD features as well as about architecture and some internals. This in-depth material will be handy for DevOps engineers who are responsible for running Argo CD for a whole organization and need to deal with scalability and performance challenges. The book provides a description of the best practices and patterns for running and managing Argo CD. I would definitely recommend it to anyone who wants to get into GitOps or who is already familiar with or looking to learn about advanced topics.

Alexander Matyushentsev

Co-founder and Chief Architect at Akuity

Contributors

About the authors

Liviu Costea started as a developer in the early 2000s and his career path led him to different roles, from developer to coding architect, and from team lead to the Chief Technical Officer. In 2012, he transitioned to DevOps when, at a small company, someone had to start working on pipelines and automation because the traditional way wasn't scalable anymore.

In 2018, he started with the platform team and then became the tech lead in the release team at Mambu, where they designed most of the **Continuous Integration/Continuous Deployment (CI/CD)** pipelines, adopting GitOps practices. They have been live with Argo CD since 2019. More recently, he joined Juni, a promising start-up, where they are planning GitOps adoption. For his contributions to OSS projects, including Argo CD, he was named a CNCF ambassador in August 2020.

Spiros Economakis started as a software engineer in 2010 and went through a series of jobs and roles, from software engineer and software architect to head of cloud. In 2013, he founded his own start-up, and that was his first encounter with DevOps culture. With a small team, he built a couple of CI/CD pipelines for a microservice architecture and mobile app releases. After this, with most of the companies he has been involved with, he has influenced DevOps culture and automation.

In 2019, he started as an SRE in Lenses (acquired by Celonis) and soon introduced the organization to Kubernetes, GitOps, and the cloud. He transitioned to a position as head of cloud, where he introduced GitOps across the whole company and used Argo CD to bootstrap K8s clusters and continuous delivery practices. Now, he works in an open source company called Mattermost as a senior engineering manager, where he transformed the old GitOps approach (`fluxcd`) to GitOps 2.0 with Argo CD and built a scalable architecture for multi-tenancy as the single GitOps platform in the company.

About the reviewer

Roel Reijerse studied electrical engineering and computer science at Delft University of Technology, with a specialization in computer graphics as part of his MSc. After several years of working as an embedded software engineer, he moved to backend engineering. Currently, he is employed by Celonis, where he works on a real-time streaming data platform managed by Argo CD.

Sai Kothapalle works as the Lead Site Reliability engineer at Enix. His experience includes working on distributed systems, running Kubernetes and ArgoCD Tools at scale for cloud providers, fintech companies and clients.

Table of Contents

Part 1: The Fundamentals of GitOps and Argo CD

1

2

Part 2: Argo CD as a Site Reliability Engineer

3

Operating Argo CD   49

4

Access Control   79

Part 3: Argo CD in Production

5

6

Preface

GitOps is not a topic that is hard to understand; you use a Git repository to declaratively define the state of your environments and by doing so, you gain versioning and changes by merge requests, which makes the whole system auditable.

But once you start adopting it and use a tool such as Argo CD, things will start becoming more complex. First, you need to set up Argo CD correctly, keeping in mind things such as observability and high availability. Then, you need to think about the CI/CD pipelines and how the new GitOps repositories will be integrated with them. And there will be organizational challenges: how do you integrate each team into this new setup? Most likely, they had different types of Kubernetes access based on the namespace they were deploying to, so **Role-based Access Control** (**RBAC**) took time to be properly configured, and now you need to take into consideration how the existing teams' access will be transferred to the new GitOps engine.

Of course, there are many resources out there (articles, videos, and courses), but it is not easy to navigate them as they only deal with parts of these topics, and not all of them have a good level of detail.

So, it is not easy to gain an idea of what the overall adoption of Argo CD means.

We wrote this book in order for you to have a guide to understand the steps you need to take to start using Argo CD, to allow you to see the complete picture, from installation to setting up proper access control, and the challenges you will face when running it in production, including advanced scenarios and troubleshooting.

We started with GitOps early at our companies and we both were able to see the journey up close. Initially, we even thought about building our own GitOps operator, (like, how hard can it be?), but after 2-3 weeks of analyzing what we needed to do, we dropped the idea. We faced many challenges, some we handled better, while some took us a lot of time to get right, but we learned from all of them, and this is what we want to share with you. We know that, by using this book, you will be able to accelerate your Argo CD and GitOps adoption.

Who this book is for

If you're a software developer, DevOps engineer, or SRE who is responsible for building CD pipelines for projects running on Kubernetes and you want to advance in your career, this book is for you. Basic knowledge of Kubernetes, Helm, or Kustomize, and CD pipelines will be useful to get the most out of this book.

What this book covers

Chapter 1, *GitOps and Kubernetes*, explores how Kubernetes made it possible to introduce the GitOps concept. We will discover its declarative APIs, and see how we can apply resources from files, folders, and, in the end, Git repositories.

Chapter 2, *Getting Started with Argo CD*, explores the core concepts of Argo CD and its architectural overview and goes through the necessary vocabulary you need to know in order to be able to deep dive into the tool.

Chapter 3, *Operating Argo CD*, covers installing Argo CD using HA manifests, going through some of the most meaningful configuration options, preparing for disaster recovery, and discovering some relevant metrics being exposed.

Chapter 4, *Access Control*, discovers how to set up user access and the options for connecting via the CLI, web UI, or a CI/CD pipeline. It goes into detail about RBAC and SSO and the different options to configure them.

Chapter 5, *Argo CD Bootstrap K8s Cluster*, shows how we can create a Kubernetes cluster using infrastructure as code and then set up the required applications with Argo CD, identifying the security challenges you will encounter when deploying the applications.

Chapter 6, *Designing Argo CD Delivery Pipelines*, continues (based on the infrastructure setup of the previous chapter) to demonstrate real deployment strategies, including dealing with secrets and getting familiarized with Argo Rollouts.

Chapter 7, *Troubleshooting Argo CD*, addresses some of the issues you will most likely encounter during installation and your day-to-day work and also takes a look at ways to improve Argo CD performance.

Chapter 8, *YAML and Kubernetes Manifests (Parsing and Verification)*, looks at the tools we can use to validate the YAML manifests we will write, to verify them with the common best practices, check against Kubernetes schemas, or even perform your own extended validations written in Rego.

Chapter 9, *Future and Conclusion*, deals with the GitOps engine and `kubernetes-sigs/cli-utils`, how it was factored out from Argo CD or the K8s community, and what the teams are trying to achieve with them – having a set of libraries to provide a set of basic GitOps features.

To get the most out of this book

To run the code from all the chapters, you will need access to a Kubernetes cluster, which can be a local one, with the exception of the HA installation, which requires a cluster with multiple nodes. The tools we will use the most are kubectl, Helm, and Kustomize. In the Kubernetes cluster, we will install Argo CD, and the instructions can be found in *Chapter 2*, *Getting Started with Argo CD* for the normal installation or *Chapter 3*, *Operating Argo CD* for the HA one.

Software/hardware covered in the book	Operating system requirements
Argo CD v2.1 and v2.2	Windows, macOS, or Linux

For some of the chapters, such as *Chapters 3, Operating Argo CD* and *Chapter 5*, Argo CD Bootstrap K8s Cluster we work with AWS EKS clusters, so you will need an AWS account set up and the AWS CLI installed. In *Chapter 3*, we also mention the `eksctl` CLI in order to ease the creation of the cluster where we will perform the HA installation, while in *Chapter 5, Argo CD Bootstrap k8s Cluster*, we recommend using Terraform for the cluster creation.

If you are using the digital version of this book, we advise you to type the code yourself or access the code from the book's GitHub repository (a link is available in the next section). Doing so will help you avoid any potential errors related to the copying and pasting of code.

Download the example code files

You can download the example code files for this book from GitHub at `https://github.com/PacktPublishing/ArgoCD-in-Practice`. If there's an update to the code, it will be updated in the GitHub repository.

We also have other code bundles from our rich catalog of books and videos available at `https://github.com/PacktPublishing/`. Check them out!

Download the color images

We also provide a PDF file that has color images of the screenshots and diagrams used in this book. You can download it here: `https://packt.link/HfXCL`.

Conventions used

There are a number of text conventions used throughout this book.

`Code in text`: Indicates code words in text, database table names, folder names, filenames, file extensions, pathnames, dummy URLs, user input, and Twitter handles. Here is an example: "Create the new file named `argocd-rbac-cm.yaml` in the same location where we have `argocd-cm.yaml`."

A block of code is set as follows:

```
apiVersion: v1
kind: ConfigMap
metadata:
  name: argocd-cm
```

```
data:
  accounts.alina: apiKey, login
```

When we wish to draw your attention to a particular part of a code block, the relevant lines or items are set in bold:

```
patchesStrategicMerge:
  - patches/argocd-cm.yaml
  - patches/argocd-rbac-cm.yaml
```

Any command-line input or output is written as follows:

```
kubectl get secret argocd-initial-admin-secret -n argocd -o
jsonpath='{.data.password}' | base64 -d
```

Bold: Indicates a new term, an important word, or words that you see onscreen. For instance, words in menus or dialog boxes appear in **bold**. Here is an example: "You can use the UI by navigating to the **User-Info** section."

> **Tips or Important Notes**
> Appear like this.

Get in touch

Feedback from our readers is always welcome.

General feedback: If you have questions about any aspect of this book, email us at customercare@packtpub.com and mention the book title in the subject of your message.

Errata: Although we have taken every care to ensure the accuracy of our content, mistakes do happen. If you have found a mistake in this book, we would be grateful if you would report this to us. Please visit www.packtpub.com/support/errata and fill in the form.

Piracy: If you come across any illegal copies of our works in any form on the internet, we would be grateful if you would provide us with the location address or website name. Please contact us at copyright@packt.com with a link to the material.

If you are interested in becoming an author: If there is a topic that you have expertise in and you are interested in either writing or contributing to a book, please visit authors.packtpub.com

Share Your Thoughts

Once you've read *Argo CD in Practice*, we'd love to hear your thoughts! Scan the QR code below to go straight to the Amazon review page for this book and share your feedback.

https://packt.link/r/180323332X

Your review is important to us and the tech community and will help us make sure we're delivering excellent quality content.

Download a free PDF copy of this book

Thanks for purchasing this book!

Do you like to read on the go but are unable to carry your print books everywhere?

Is your eBook purchase not compatible with the device of your choice?

Don't worry, now with every Packt book you get a DRM-free PDF version of that book at no cost.

Read anywhere, any place, on any device. Search, copy, and paste code from your favorite technical books directly into your application.

The perks don't stop there, you can get exclusive access to discounts, newsletters, and great free content in your inbox daily

Follow these simple steps to get the benefits:

1. Scan the QR code or visit the link below

https://packt.link/free-ebook/9781803233321

2. Submit your proof of purchase

3. That's it! We'll send your free PDF and other benefits to your email directly

Part 1: The Fundamentals of GitOps and Argo CD

This part serves as an introduction to GitOps as a practice and will cover the advantages of using it.

This part of the book comprises the following chapters:

- *Chapter 1, GitOps and Kubernetes*
- *Chapter 2, Getting Started with Argo CD*

1
GitOps and Kubernetes

In this chapter, we're going to see what GitOps is and how the idea makes a lot of sense in a Kubernetes cluster. We will get introduced to specific components, such as the **application programming interface** (**API**) server and controller manager that make the cluster react to state changes. We will start with imperative APIs and get through the declarative ones and will see how applying a file and a folder up to applying a Git repository was just one step—and, when it was taken, GitOps appeared.

We will cover the following main topics in this chapter:

- What is GitOps?

- Kubernetes and GitOps

- Imperative and declarative APIs

- Building a simple GitOps operator

- **Infrastructure as code** (**IaC**) and GitOps

Technical requirements

For this chapter, you will need access to a Kubernetes cluster, and a local one such as `minikube` (`https://minikube.sigs.k8s.io/docs/`) or `kind` (`https://kind.sigs.k8s.io`) will do. We are going to interact with the cluster and send commands to it, so you also need to have `kubectl` installed (`https://kubernetes.io/docs/tasks/tools/#kubectl`).

We are going to write some code, so a code editor will be needed. I am using **Visual Studio Code** (**VS Code**) (`https://code.visualstudio.com`), and we are going to use the Go language, which needs installation too: `https://golang.org` (the current version of Go is `1.16.7`; the code should work with it). The code can be found at `https://github.com/PacktPublishing/ArgoCD-in-Practice` in the `ch01` folder.

What is GitOps?

The term *GitOps* was coined back in 2017 by people from Weaveworks, who are also the authors of a GitOps tool called Flux. Since then, I have seen how GitOps turned into a buzzword, up to being named the next important thing after **development-operations (DevOps)**. If you search for definitions and explanations, you will find a lot of them: it has been defined as operations via **pull requests (PRs)** (`https://www.weave.works/blog/gitops-operations-by-pull-request`) or taking development practices (version control, collaboration, compliance, **continuous integration/ continuous deployment (CI/CD)**) and applying them to infrastructure automation (`https:// about.gitlab.com/topics/gitops/`).

Still, I think there is one definition that stands out. I am referring to the one created by the GitOps Working Group (`https://github.com/gitops-working-group/gitops-working-group`), which is part of the **Application Delivery Technical Advisory Group (Application Delivery TAG)** from the **Cloud Native Computing Foundation (CNCF)**. The Application Delivery TAG is specialized in building, deploying, managing, and operating cloud-native applications (`https:// github.com/cncf/tag-app-delivery`). The workgroup is made up of people from different companies with the purpose of building a vendor-neutral, principle-led definition for GitOps, so I think these are good reasons to take a closer look at their work.

The definition is focused on the principles of GitOps, and five are identified so far (this is still a draft), as follows:

- Declarative configuration
- Version-controlled immutable storage
- Automated delivery
- Software agents
- Closed loop

It starts with **declarative configuration**, which means we want to express our intent, an end state, and not specific actions to execute. It is not an imperative style where you say, "*Let's start three more containers,*" but instead, you declare that you want to have three containers for this application, and an agent will take care of reaching that number, which might mean it needs to stop two running containers if there are five up right now.

Git is being referred to here as **version-controlled and immutable storage**, which is fair because while it is the most used source control system right now, it is not the only one, and we could implement GitOps with other source control systems.

Automated delivery means that we shouldn't have any manual actions once the changes reach the **version control system (VCS)**. After the configuration is updated, it comes to **software agents** to make sure that the necessary actions to reach the new declared configuration are being taken. Because we are expressing the desired state, the actions to reach it need to be calculated. They result from the difference between the actual state of the system and the desired state from the version control—and this is what the **closed loop** part is trying to say.

While GitOps originated in the Kubernetes world, this definition is trying to take that out of the picture and bring the preceding principles to the whole software world. In our case, it is still interesting to see what made GitOps possible and dive a little bit deeper into what those software agents are in Kubernetes or how the closed loop is working here.

Kubernetes and GitOps

It is hard not to hear about Kubernetes these days—it is probably one of the most well-known open source projects at the moment. It originated somewhere around 2014 when a group of engineers from Google started building a container orchestrator based on the experience they accumulated working with Google's own internal orchestrator named Borg. The project was open sourced in 2014 and reached its 1.0.0 version in 2015, a milestone that encouraged many companies to take a closer look at it.

Another reason that led to its fast and enthusiastic adoption by the community is the governance of CNCF (`https://www.cncf.io`). After making the project open source, Google started discussing with the Linux Foundation (`https://www.linuxfoundation.org`) creating a new nonprofit organization that would lead the adoption of open source cloud-native technologies. That's how CNCF came to be created while Kubernetes became its seed project and *KubeCon* its major developer conference. When I said CNCF governance, I am referring mostly to the fact that every project or organization inside CNCF has a well-established structure of maintainers and details how they are nominated, how decisions are taken in these groups, and that no company can have a simple majority. This ensures that no decision will be taken without community involvement and that the overall community has an important role to play in a project life cycle.

Architecture

Kubernetes has become so big and extensible that it is really hard to define it without using abstractions such as *a platform for building platforms*. This is because it is just a starting point—you get many pieces, but you have to put them together in a way that works for you (and GitOps is one of those pieces). If we say that it is a container orchestration platform, this is not entirely true because you can also run **virtual machines** (**VMs**) with it, not just containers (for more details, please check `https://ubuntu.com/blog/what-is-kata-containers`); still, the orchestration part remains true.

Its components are split into two main parts—first is the *control plane*, which is made of a **REpresentational State Transfer** (**REST**) API server with a database for storage (usually `etcd`), a controller manager used to run multiple control loops, a scheduler that has the job of assigning a node for our Pods (a Pod is a logical grouping of containers that helps to run them on the same node—find out more at `https://kubernetes.io/docs/concepts/workloads/pods/`), and a cloud controller manager to handle any cloud-specific work. The second piece is the data plane, and while the control plane is about managing the cluster, this one is about what happens on the nodes running the user workloads. A node that is part of a Kubernetes cluster will have a container runtime (which can be Docker, CRI-O, or `containerd`, and there are a few others), `kubelet`, which takes care of the connection between the REST API server and the container runtime of the node, and `kube-proxy`, responsible for abstracting the network at the node level. See the next diagram for details of how all the components work together and the central role played by the API server.

We are not going to enter into the details of all these components; instead, for us, the REST API server that makes the declarative part possible and the controller manager that makes the system converge to the desired state are important, so we want to dissect them a little bit.

The following diagram shows an overview of a typical Kubernetes architecture:

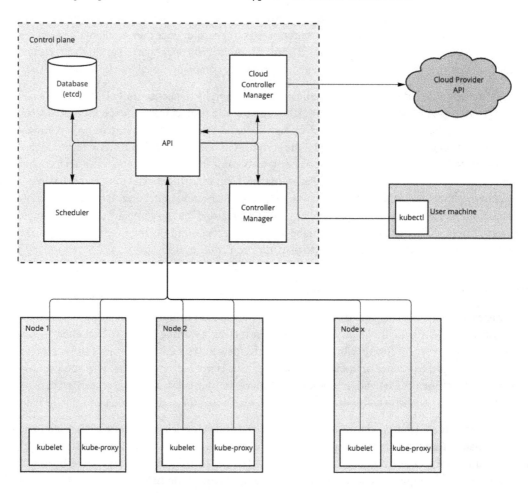

Figure 1.1 – Kubernetes architecture

> **Note**
>
> When looking at an architecture diagram, you need to know that it is only able to catch a part of the whole picture. For example, here, it seems that the cloud provider with its API is an external system, but actually, all the nodes and the control plane are created in that cloud provider.

HTTP REST API server

Viewing Kubernetes from the perspective of the **HyperText Transfer Protocol** (HTTP) REST API server makes it like any classic application with REST endpoints and a database for storing state—in our case, usually `etcd`—and with multiple replicas of the web server for **high availability** (HA). What is important to emphasize is that anything we want to do with Kubernetes we need to do via the API; we can't connect directly to any other component, and this is true also for the internal ones: they can't talk directly between them—they need to go through the API.

From our client machines, we don't query the API directly (such as by using `curl`), but instead, we use this `kubectl` client application that hides some of the complexity, such as authentication headers, preparing request content, parsing the response body, and so on.

Whenever we do a command such as `kubectl get pods`, there is an **HTTP Secure** (**HTTPS**) call to the API server. Then, the server goes to the database to fetch details about the Pods, and a response is created and pushed back to the client. The `kubectl` client application receives it, parses it, and is able to display a nice output suited to a human reader. In order to see what exactly happens, we can use the verbose global flag of `kubectl` (`--v`), for which the higher value we set, the more details we get.

For an exercise, do try `kubectl get pods --v=6`, when it just shows that a `GET` request is performed, and keep increasing `--v` to 7, 8, 9, and more so that you will see the HTTP request headers, the response headers, part or all of the **JavaScript Object Notation** (**JSON**) response, and many other details.

The API server itself is not responsible for actually changing the state of the cluster—it updates the database with the new values and, based on such updates, other things are happening. The actual state changes are done by controllers and components such as `scheduler` or `kubelet`. We are going to drill down into controllers as they are important for our GitOps understanding.

Controller manager

When reading about Kubernetes (or maybe listening to a podcast), you will hear the word *controller* quite often. The idea behind it comes from industrial automation or robots, and it is about the converging control loop.

Let's say we have a robotic arm and we give it a simple command to move at a 90-degree position. The first thing that it will do is to analyze its current state; maybe it is already at 90 degrees and there is nothing to do. If it isn't in the right position, the next thing is to calculate the actions to take in order to get to that position, and then, it will try to apply those actions to reach its relative place.

We start with the *observe* phase, where we compare the desired state with the current state, then we have the *diff* phase, where we calculate the actions to apply, and in the *action* phase, we perform those actions. And again, after we perform the actions, it starts the observe phase to see if it is in the right position; if not (maybe something blocked it from getting there), actions are calculated, and we get into applying the actions, and so on until it reaches the position or maybe runs out of battery or something. This control loop continues on and on until in the observe phase, the current state matches the desired state, so there will be no actions to calculate and apply. You can see a representation of the process in the following diagram:

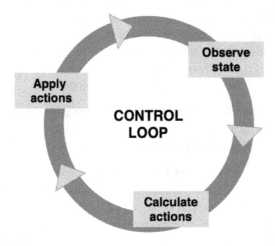

Figure 1.2 – Control loop

In Kubernetes, there are many controllers. We have the following:

- **ReplicaSet**: https://kubernetes.io/docs/concepts/workloads/controllers/replicaset/

- **HorizontalPodAutoscaler** (**HPA**): https://kubernetes.io/docs/tasks/run-application/horizontal-pod-autoscale/

- And a few others can be found here, but this isn't a complete list: https://kubernetes.io/docs/concepts/workloads/controllers/

The ReplicaSet controller is responsible for running a fixed number of Pods. You create it via kubectl and ask to run three instances, which is the desired state. So, it starts by checking the current state: how many Pods we have running right now; it calculates the actions to take: how many more Pods to start or terminate in order to have three instances; it then performs those actions. There is also the HPA controller, which, based on some metrics, is able to increase or decrease the number of Pods for a Deployment (a Deployment is a construct built on top of Pods and ReplicaSets that allows us to define ways to update Pods (https://kubernetes.io/docs/concepts/workloads/

`controllers/deployment/`)), and a Deployment relies on a ReplicaSet controller it builds internally in order to update the number of Pods. After the number is modified, it is still the ReplicaSet controller that runs the control loop to reach the number of desired Pods.

The controller's job is to make sure that the actual state matches the desired state, and they never stop trying to reach that final state. And, more than that, they are specialized in types of resources—each takes care of a small piece of the cluster.

In the preceding examples, we talked about internal Kubernetes controllers, but we can also write our own, and that's what Argo CD really is—a controller, its control loop taking care that the state declared in a Git repository matches the state from the cluster. Well, actually, to be correct, it is not a controller but an operator, the difference being that controllers work with internal Kubernetes objects while operators deal with two domains: Kubernetes and something else. In our case, the Git repository is the outside part handled by the operator, and it does that using something called **custom resources**, a way to extend Kubernetes functionality (`https://kubernetes.io/docs/concepts/extend-kubernetes/api-extension/custom-resources/`).

So far, we have looked at the Kubernetes architecture with the API server connecting all the components and how the controllers are always working within control loops to get the cluster to the desired state. Next, we will get into details on how we can define the desired state: we will start with the imperative way, continue with the more important declarative way, and show how all these get us one step closer to GitOps.

Imperative and declarative APIs

We discussed a little bit about the differences between an imperative style where you clearly specify actions to take—such as *start three more Pods*—and a declarative one where you specify your intent—such as *there should be three Pods running for the deployment*—and actions need to be calculated (you might increase or decrease the Pods or do nothing if three are already running). Both imperative and declarative ways are implemented in the `kubectl` client.

Imperative – direct commands

Whenever we create, update, or delete a Kubernetes object, we can do it in an imperative style.

To create a namespace, run the following command:

```
kubectl create namespace test-imperative
```

Then, in order to see the created namespace, use the following command:

```
kubectl get namespace test-imperative
```

Create a deployment inside that namespace, like so:

```
kubectl create deployment nginx-imperative --image=nginx -n
test-imperative
```

Then, you can use the following command to see the created deployment:

```
kubectl get deployment -n test-imperative nginx-imperative
```

To update any of the resources we created, we can use specific commands, such as `kubectl label` to modify the resource labels, `kubectl scale` to modify the number of Pods in a Deployment, ReplicaSet, `StatefulSet`, or `kubectl set` for changes such as environment variables (`kubectl set env`), container images (`kubectl set image`), resources for a container (`kubectl set resources`), and a few more.

If you want to add a label to the namespace, you can run the following command:

```
kubectl label namespace test-imperative namespace=imperative-
apps
```

In the end, you can remove objects created previously with the following commands:

```
kubectl delete deployment -n test-imperative nginx-imperative
kubectl delete namespace test-imperative
```

Imperative commands are clear on what they do, and it makes sense when you use them for small objects such as namespaces. But for more complex ones, such as Deployments, we can end up passing a lot of flags to it, such as specifying a container image, image tag, pull policy, if a secret is linked to a pull (for private image registries), and the same for `init` containers and many other options. Next, let's see if there are better ways to handle such a multitude of possible flags.

Imperative – with config files

Imperative commands can also make use of configuration files, which make things easier because they significantly reduce the number of flags we would need to pass to an imperative command. We can use a file to say what we want to create.

This is what a namespace configuration file looks like—the simplest version possible (without any labels or annotations). The following files can also be found at https://github.com/PacktPublishing/ArgoCD-in-Practice/tree/main/ch01/imperative-confi

Copy the following content into a file called `namespace.yaml`:

```
apiVersion: v1
kind: Namespace
metadata:
  name: imperative-config-test
```

Then, run the following command:

```
kubectl create -f namespace.yaml
```

Copy the following content and save it in a file called `deployment.yaml`:

```
apiVersion: apps/v1
kind: Deployment
metadata:
  name: nginx-deployment
  namespace: imperative-config-test
spec:
  selector:
    matchLabels:
      app: nginx
  template:
    metadata:
      labels:
        app: nginx
    spec:
      containers:
      - name: nginx
        image: nginx
```

Then, run the following command:

```
kubectl create -f deployment.yaml
```

By running the preceding commands, we create one namespace and one Deployment, similar to what we have done with imperative direct commands. You can see this is easier than passing all the flags to `kubectl create deployment`. What's more, not all the fields are available as flags, so using a configuration file can become mandatory in many cases.

We can also modify objects via the config file. Here is an example of how to add labels to a namespace. Update the namespace we used before with the following content (notice the extra two rows starting with `labels`). The updated namespace can be seen in the official `https://github.com/PacktPublishing/ArgoCD-in-Practice/tree/main/ch01/imperative-config` repository in the `namespace-with-labels.yaml` file:

```
apiVersion: v1
kind: Namespace
metadata:
  name: imperative-config-test
  labels:
    name: imperative-config-test
```

And then, we can run the following command:

```
kubectl replace -f namespace.yaml
```

And then, to see if a label was added, run the following command:

```
kubectl get namespace imperative-config-test -o yaml
```

This is a good improvement compared to passing all the flags to the commands, and it makes it possible to store those files in version control for future reference. Still, you need to specify your intention if the resource is new, so you use `kubectl create`, while if it exists, you use `kubectl replace`. There are also some limitations: the `kubectl replace` command performs a full object update, so if someone modified something else in between (such as adding an annotation to the namespace), those changes will be lost.

Declarative – with config files

We just saw how easy it is to use a config file to create something, so it would be great if we could modify the file and just call some `update/sync` command on it. We could modify the labels inside the file instead of using `kubectl label` and could do the same for other changes, such as scaling the Pods of a Deployment, setting container resources, container images, and so on. And there is such a command that you can pass any file to it, new or modified, and it will be able to make the right adjustments to the API server: `kubectl apply`.

Please create a new folder called `declarative-files` and place the `namespace.yaml` file in it, with the following content (the files can also be found at `https://github.com/PacktPublishing/ArgoCD-in-Practice/tree/main/ch01/declarative-files`):

```
apiVersion: v1
kind: Namespace
```

```
metadata:
  name: declarative-files
```

Then, run the following command:

```
kubectl apply -f declarative-files/namespace.yaml
```

The console output should then look like this:

```
namespace/declarative-files created
```

Next, we can modify the namespace.yaml file and add a label to it directly in the file, like so:

```
apiVersion: v1
kind: Namespace
metadata:
  name: declarative-files
  labels:
    namespace: declarative-files
```

Then, run the following command again:

```
kubectl apply -f declarative-files/namespace.yaml
```

The console output should then look like this:

```
namespace/declarative-files configured
```

What happened in both of the preceding cases? Before running any command, our client (or our server—there is a note further on in this chapter explaining when client-side or server-side apply is used) compared the existing state from the cluster with the desired one from the file, and it was able to calculate the actions that needed to be applied in order to reach the desired state. In the first apply example, it realized that the namespace didn't exist and it needed to create it, while in the second one, it found that the namespace exists but it didn't have a label, so it added one.

Next, let's add the Deployment in its own file called deployment.yaml in the same declarative-files folder, as follows:

```
apiVersion: apps/v1
kind: Deployment
metadata:
  name: nginx
  namespace: declarative-files
```

```
spec:
  selector:
    matchLabels:
      app: nginx
  template:
    metadata:
      labels:
        app: nginx
    spec:
      containers:
      - name: nginx
        image: nginx
```

And we will run the following command that will create a Deployment in the namespace:

```
kubectl apply -f declarative-files/deployment.yaml
```

If you want, you can make the changes to the deployment.yaml file (labels, container resources, images, environment variables, and so on) and then run the kubectl apply command (the complete one is the preceding one), and the changes you made will be applied to the cluster.

Declarative – with config folder

In this section, we will create a new folder called declarative-folder and two files inside of it.

Here is the content of the namespace.yaml file (the code can also be found here: https://github.com/PacktPublishing/ArgoCD-in-Practice/tree/main/ch01/declarative-folder):

```
apiVersion: v1
kind: Namespace
metadata:
  name: declarative-folder
```

Here is the content of the deployment.yaml file:

```
apiVersion: apps/v1
kind: Deployment
metadata:
 name: nginx
  namespace: declarative-folder
```

```
spec:
  selector:
    matchLabels:
      app: nginx
  template:
    metadata:
      labels:
        app: nginx
    spec:
      containers:
      - name: nginx
        image: nginx
```

And then, we will run the following command:

```
kubectl apply -f declarative-folder
```

Most likely, you will see the following error, which is expected, so don't worry:

```
namespace/declarative-folder created
Error from server (NotFound): error when creating "declarative-
folder/deployment.yaml": namespaces "declarative-folder" not
found
```

That is because those two resources are created at the same time, but deployment depends on the namespace, so when a deployment needs to be created, it needs to have the namespace ready. We see the message says that a namespace was created but the API calls were done at the same time and on the server, so the namespace was not available when the deployment started its creation flow. We can fix this by running the following command again:

```
kubectl apply -f declarative-folder
```

And in the console, we should see the following output:

```
deployment.apps/nginx created
namespace/declarative-folder unchanged
```

Because the namespace already existed, it was able to create a deployment inside it while no change was made to the namespace.

The kubectl apply command took the whole content of the declarative-folder folder, made the calculations for each resource found in those files, and then called the API server with the changes. We can apply entire folders, not just files, though it can get trickier if the resources depend on each other, and we can modify those files and call the `apply` command for the folder, and the changes will get applied. Now, if this is how we build applications in our clusters, then we had better save all those files in source control for future reference so that it will get easier to apply changes after some time.

But what if we could apply a Git repository directly, not just folders and files? After all, a local Git repository is a folder, and in the end, that's what a GitOps operator is: a `kubectl apply` command that knows how to work with Git repositories.

> **Note**
>
> The `apply` command was initially implemented completely on the client side. This means the logic for finding changes was running on the client, and then specific imperative APIs were called on the server. But more recently, the `apply` logic moved to the server side; all objects have an `apply` method (from a REST API perspective, it is a `PATCH` method with an `application/apply-patch+yaml` content-type header), and it is enabled by default starting with version 1.16 (more on the subject here: `https://kubernetes.io/docs/reference/using-api/server-side-apply/`).

Building a simple GitOps operator

Now that we have seen how the control loop works, have experimented with declarative commands, and know how to work with basic Git commands, we have enough information to build a basic GitOps operator. We now need three things created, as follows:

- We will initially clone a Git repository and then pull from it to keep it in sync with remote.
- We'll take what we found in the Git repository and try to apply it.
- We'll do this in a loop so that we can make changes to the Git repository and they will be applied.

The code is in Go; this is a newer language from Google, and many **operations (ops)** tools are built with it, such as Docker, Terraform, Kubernetes, and Argo CD.

> **Note**
>
> For real-life controllers and operators, certain frameworks should be used, such as the Operator Framework (`https://operatorframework.io`), Kubebuilder (`https://book.kubebuilder.io`), or sample-controller (`https://github.com/kubernetes/sample-controller`).

All the code for our implementation can be found at `https://github.com/PacktPublishing/ArgoCD-in-Practice/tree/main/ch01/basic-gitops-operator`, while the **YAML**

Ain't Markup Language (**YAML**) manifests we will be applying are at `https://github.com/PacktPublishing/ArgoCD-in-Practice/tree/main/ch01/basic-gitops-operator-config`.

The `syncRepo` function receives the repository **Uniform Resource Locator** (**URL**) to clone and keep in sync, as well as the local path where to do it. It then tries to clone the repository using a function from the `go-git` library (`https://github.com/go-git/go-git`), `git.PlainClone`. If it fails with a `git.ErrRepositoryAlreadyExists` error, this means we have already cloned the repository and we need to pull it from the remote to get the latest updates. And that's what we do next: we open the Git repository locally, load the worktree, and then call the `Pull` method. This method can give an error if everything is up to date and there is nothing to download from the remote, so for us, this case is normal (this is the condition: `if err != nil && err == git.NoErrAlreadyUpToDate`). The code is illustrated in the following snippet:

```go
func syncRepo(repoUrl, localPath string) error {
    _, err := git.PlainClone(localPath, false, &git.
CloneOptions{
        URL:      repoUrl,
        Progress: os.Stdout,
    })

    if err == git.ErrRepositoryAlreadyExists {
        repo, err := git.PlainOpen(localPath)
        if err != nil {
            return err
        }
        w, err := repo.Worktree()
        if err != nil {
            return err
        }
        err = w.Pull(&git.PullOptions{
            RemoteName: "origin",
            Progress:   os.Stdout,
        })
        if err == git.NoErrAlreadyUpToDate {
            return nil
        }
        return err

    }
```

```
        return err
    }
```

Next, inside the `applyManifestsClient` method, we have the part where we apply the content of a folder from the repository we downloaded. Here, we create a simple wrapper over the `kubectl apply` command, passing as a parameter the folder where the YAML manifests are from the repository we cloned. Instead of using the `kubectl apply` command, we can use the Kubernetes APIs with the `PATCH` method (with the `application/apply-patch+yaml` content-type header), which means calling `apply` on the server side directly. But it complicates the code, as each file from the folder needs to be read and transformed into its corresponding Kubernetes object in order to be able to pass it as a parameter to the API call. The `kubectl apply` command does this already, so this was the simplest implementation possible. The code is illustrated in the following snippet:

```go
func applyManifestsClient(localPath string) error {
    dir, err := os.Getwd()
    if err != nil {
        return err
    }
    cmd := exec.Command("kubectl", "apply", "-f", path.Join(dir,
localPath))
    cmd.Stdout = os.Stdout
    cmd.Stderr = os.Stderr
    err = cmd.Run()
    return err
}
```

Finally, the `main` function is from where we call these functionalities, sync the Git repository, apply manifests to the cluster, and do it in a loop at a 5-second interval (I went with a short interval for demonstration purposes; in live scenarios, Argo CD—for example—does this synchronization every 3 minutes). We define the variables we need, including the Git repository we want to clone, so if you will fork it, please update the `gitopsRepo` value. Next, we call the `syncRepo` method, check for any errors, and if all is good, we continue by calling `applyManifestsClient`. The last rows are how a timer is implemented in Go, using a channel.

Note: Complete code file

For a better overview, we also add the `package` and `import` declaration; this is the complete implementation that you can copy into the `main.go` file.

Here is the code for the `main` function where everything is put together:

```go
package main
import (
    "fmt"
    "os"
    "os/exec"
    "path"
    "time"
    "github.com/go-git/go-git/v5"
)
func main() {
    timerSec := 5 * time.Second

    gitopsRepo := "https://github.com/PacktPublishing/ArgoCD-in-
Practice.git"    localPath := "tmp/"
    pathToApply := "ch01/basic-gitops-operator-config"
    for {
        fmt.Println("start repo sync")
        err := syncRepo(gitopsRepo, localPath)
        if err != nil {
            fmt.Printf("repo sync error: %s", err)
            return
        }
        fmt.Println("start manifests apply")
        err = applyManifestsClient(path.Join(localPath,
pathToApply))
        if err != nil {
            fmt.Printf("manifests apply error: %s", err)
        }
        syncTimer := time.NewTimer(timerSec)
        fmt.Printf("\n next sync in %s \n", timerSec)
        <-syncTimer.C
    }
}
```

To make the preceding code work, go to a folder and run the following command (just replace <your-username>):

```
go mod init github.com/<your-username>/basic-gitops-operator
```

This creates a go.mod file where we will store the Go modules we need. Then, create a file called main.go and copy the preceding pieces of code in it, and the three functions syncRepo, applyManifestsClient, and main (also add the package and import declarations that come with the main function). Then, run the following command:

```
go get .
```

This will download all the modules (don't miss the last dot).

And the last step is to actually execute everything we put together with the following command:

```
go run main.go
```

Once the application starts running, you will notice a tmp folder created, and inside it, you will find the manifests to be applied to the cluster. The console output should look something like this:

```
start repo sync
Enumerating objects: 36, done.
Counting objects: 100% (36/36), done.
Compressing objects: 100% (24/24), done.
Total 36 (delta 8), reused 34 (delta 6), pack-reused 0
start manifests apply
namespace/nginx created
Error from server (NotFound): error when creating "<>/argocd-
in-practice/ch01/basic-gitops-operator/tmp/ch01/basic-gitops-
operator-config/deployment.yaml": namespaces "nginx" not found
manifests apply error: exit status 1
next sync in 30s
start repo sync
start manifests apply
deployment.apps/nginx created
namespace/nginx unchanged
```

You can see the same error since, as we tried applying an entire folder, this is happening now too, but on the operator's second run, the deployment is created successfully. If you look in your cluster, you should find a namespace called nginx and, inside it, a deployment also called nginx. Feel free to fork the repository and make changes to the operator and to the config it is applying.

> **Note: Apply namespace first**
>
> The problem with namespace creation was solved in Argo CD by identifying them and applying namespaces first.

We created a simple GitOps operator, showing the steps of cloning and keeping the Git repository in sync with the remote and taking the contents of the repository and applying them. If there was no change to the manifests, then the `kubectl apply` command had nothing to modify in the cluster, and we did all this in a loop that imitates pretty closely the control loop we introduced earlier in the chapter. As a principle, this is alsowhat happens in the Argo CD implementation, but at a much higher scale and performance and with a lot of features added.

IaC and GitOps

You can find a lot of articles and blog posts trying to make comparisons between IaC and GitOps to cover the differences and, usually, how GitOps builds upon IaC principles. I would say that they have a lot of things in common—they are very similar practices that use source control for storing the state. When you say IaC these days, you are referring to practices where infrastructure is created through automation and not manually, and the infrastructure is saved as code in source control just like application code.

With IaC, you expect the changes to be applied using pipelines, a big advantage over going and starting to provision things manually. This allows us to create the same environments every time we need them, reducing the number of inconsistencies between staging and production, for example, which will translate into developers spending less time debugging special situations and problems caused by configuration drift.

The way of applying changes can be both imperative and declarative; most tools support both ways, while some are only declarative in nature (such as Terraform or CloudFormation). Initially, some started as imperative but adopted declarative configuration as it gained more traction recently (see `https://next.redhat.com/2017/07/24/ansible-declares-declarative-intent/`).

Having your infrastructure in source control adds the benefit of using PRs that will be peer-reviewed, a process that generates discussions, ideas, and improvements until changes are approved and merged. It also makes our infrastructure changes clear to everyone and auditable.

We went through all these principles when we discussed the GitOps definition created by the Application Delivery TAG at the beginning of this chapter. But more importantly, there were some more in the GitOps definition that are not part of the IaC one, such as software agents or closed loops. IaC is usually applied with a CI/CD system, resulting in a push mode whereby your pipeline connects to your system (cloud, database cluster, VM, and so on) and performs the changes. GitOps, on the other hand, is about agents that are working to reconcile the state of the system with the one declared in the source control. There is a loop where the differences are calculated and applied until the state matches. And we saw how this reconciliation happens again and again until there are no more differences discovered, this being the actual loop.

This does not happen in an IaC setup; there are no operators/controllers when talking about applying infrastructure changes. The updates are done with a push mode, which means the GitOps pull way is better in terms of security, as it is not the pipeline that has the production credentials, but your agent stores them, and it can run in the same account as your production—or at least in a separate, but trusted one.

Having agents applying your changes also means GitOps can only support the declarative way. We need to be able to specify what is the state we want to reach; we will not have too much control over how to do it as we offload that burden onto our controllers and operators.

Can a tool that was previously defined as IaC be applied in a GitOps manner? Yes, and I think we have a good example with Terraform and Atlantis (`https://www.runatlantis.io`). This is a way of running an agent (that would be Atlantis) in a remote setup, so all commands will not be executed from the pipeline, but by the agent. This means it does fit the GitOps definition, though if we go into details, we might find some mismatches regarding the closed loop.

In my opinion, Atlantis applies infrastructure changes in a GitOps way, while if you apply Terraform from your pipeline, that is IaC.

So, we don't have too many differences between these practices—they are more closely related than different. Both have the state stored in source control and open the path for making changes with PRs. In terms of differences, GitOps comes with the idea of agents and the control loop, which improves security and can only be declarative.

Summary

In this chapter, we discovered what GitOps means and which parts of Kubernetes make it possible. We checked how the API server connects everything and how controllers work, introduced a few of them, and explained how they react to state changes in an endless control loop. We took a closer look at Kubernetes' declarative nature, starting from imperative commands, then opening the path of not just applying a folder but a Git repository. In the end, we implemented a very simple controller so that you could have an idea of what Argo CD does.

In the next chapter, we are going to start exploring Argo CD, how it works, its concepts and architecture, and details around synchronization principles.

Further reading

- Kubernetes controllers architecture:

  `https://kubernetes.io/docs/concepts/architecture/controller/`

- We used `kubectl apply` in order to make changes to the cluster, but if you want to see how to use Kubernetes API from Go code, here are some examples:

  `https://github.com/kubernetes/client-go/tree/master/examples`

- More on `kubectl` declarative config options can be found at the following link:

  `https://kubernetes.io/docs/tasks/manage-kubernetes-objects/declarative-config/`

- GitOps Working Group presents GitOps principles as OpenGitOps:

  `https://opengitops.dev`

2
Getting Started with Argo CD

We will start this chapter by explaining what Argo CD is and the underlying technology that the platform is built on so that we can set the fundamentals. We will explain the core concepts of Argo CD, and we will go through the necessary vocabulary you need to know before the deep dive into it.

Then, we will describe the architectural overview of Argo CD and a typical workflow in terms of GitOps. We will describe in detail each of the core components and their responsibilities so that we will be able to understand and troubleshoot potential issues.

In the end, we will install Argo CD in the Kubernetes cluster on our local machine, and we will try to deploy an application using it and observe the GitOps phases through Argo CD.

In this chapter, we're going to cover the following main topics:

- What is Argo CD?
- Core concepts and vocabulary
- Explaining the architecture
- Synchronization principles

Technical requirements

For this chapter, you will need access to a Kubernetes cluster, which we will run locally with one of the following tools:

- Kind: `https://kind.sigs.Kubernetes.io/`
- K3s: `https://rancher.com/docs/k3s/latest/en/`
- minikube: `https://minikube.sigs.Kubernetes.io/docs/`
- MicroKubernetes: `https://microKubernetes.io/docs`

We are going to use a Helm chart to deploy Argo CD and a demo application in the local cluster. We need to install Helm CLI (`https://helm.sh/docs/intro/quickstart/`) on a local machine. We will use Visual Studio Code (`https://code.visualstudio.com`) to edit the Helm chart values. The code can be found at `https://github.com/PacktPublishing/ArgoCD-in-Practice` in the `ch02` folder.

What is Argo CD?

For several years, most of us have been using the same separate types of environments in an application, which are divided into development, test, staging, and production. Representing these in Kubernetes differs in many ways and depends on many other factors, such as the size of a team and budgeting. One of these could be different clusters per environment, or another one a separation within one cluster by namespaces. In any of these, we usually create a new namespace for the application with the necessary deployment resources and add whatever is needed to configure our application for an environment (ConfigMap, Secrets, Ingress, and so on).

The drawback with the aforementioned approach is that there will be configuration drift over time. For example, our development cluster or namespace will have the latest development version of the application or changes in Kubernetes resources such as network policies, but we will need to manually apply all the changes to the rest of the environments. A simple solution to solve this problem is to use a package manager such as Helm, Kustomize, or jsonnet so that we can define the resources of the application in a repeatable manner and as a single point of authority. For example, with Helm, we can create a couple of different versions and deploy each of them accordingly to each environment, but again, this is hard to keep track of and adds extra complexity.

But what if we were following the GitOps approach? The whole configuration will *live* in a Git repository, which will be the source of truth with pull requests and reviews for any change. Lastly, what if we had a similar controller as that described in *Chapter 1, GitOps and Kubernetes*? The controller ideally would automatically apply all of the configurations from our Git repository. Every time there is a manual change in Kubernetes resources and the desired state, which is located in the Git repository and won't match, the controller will try to reapply the desired state in order to always have the Git repository as the source of truth.

Getting familiar with Argo CD

What we described previously is called GitOps, and it's even more powerful with Argo CD. Argo CD is a declarative, GitOps continuous delivery tool for Kubernetes. One of the core components of Argo CD is called application controller, which in practice observes continuously the running applications and compares the current application state against the desired target state, for which the source of truth is a Git repository. The following uses cases are empowered with Argo CD:

- **Automated deployment**: After any Git commit action, or CI pipeline run and a manual sync trigger, the Argo CD controller will push the desired state from the Git repository into the cluster in an automated manner. This is achieved while using the Argo Events automation framework (`https://argoproj.github.io/argo-events/`) or a manual user request.

- **Observability**: Argo CD offers a UI, CLI, and Argo CD Notifications engine (`https://argocd-notifications.readthedocs.io/en/stable/`) with which we can identify whether the state of the application state is in sync with the desired state in Git.

- **Multi-tenancy**: The ability to manage and deploy to multiple clusters with **Role Based Access Control** (**RBAC**) policies for authorization.

The Argo project is a *family* of many tools, as you saw previously, which includes the following:

- Argo CD (`https://argoproj.github.io/cd`)
- Argo Rollouts (`https://argoproj.github.io/rollouts`)
- Argo Events (`https://argoproj.github.io/events`)
- Argo Workflows (`https://argoproj.github.io/workflows`)

All of them are components are complementary to each other which form some of the great GitOps and DevOps cultures and practices.

Core concepts and vocabulary

In this section, we will describe some of the core components of Argo CD, such as reconciliation, and describe in detail the core objects' **Custom Resource Definitions** (**CRDs**) of Argo CD. In parallel, we will set up a vocabulary so that we can have a common language for Argo CD operations. Lastly, we will observe the reconciliation loop and how Argo CD works.

Argo CD reconciliation

Argo CD, as described previously, has a duty to match the desired state described in a Git repository with the live state in the cluster and deliver it in the environment of our preference. This is called reconciliation, and Argo CD is in a reconciling loop from the Git repository to Kubernetes, as the following diagram shows, assuming we use Helm:

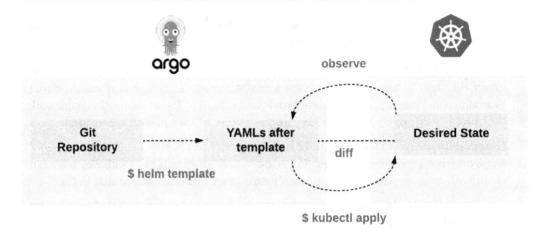

Figure 2.1 – The reconciling loop

As we see in *Figure 2.1*, Argo CD watches the Git repository and runs first a Helm template to generate the Kubernetes manifest YAML and compares them with the desired state in the cluster, which is called **sync status**. If Argo CD identifies any differences, then the templated files will be applied with `kubectl apply` and change accordingly the Kubernetes desired state, which can be automated or done manually. Also, Argo CD watches the live Kubernetes objects and compares them to the Kubernetes desired state, which is called the health status of the application.

A nice observation here is that Argo CD didn't use `helm install` but the standard `kubectl apply` instead. The reason for this is that Argo CD supports many templating tools, and its responsibility is to deploy the desired state as a GitOps declarative tool without being a wrapper to any of these tools.

Vocabulary

After introducing the GitOps and Argo CD core concepts, we need to define some required vocabulary terms that we have already used and will be using across the book:

- **Application**: A group of Kubernetes resources that are described by manifest. These are defined in Argo CD as **Custom Resource Definitions (CRDs)**: `https://kubernetes.io/docs/concepts/extend-kubernetes/api-extension/custom-resources/#customresourcedefinitions`.

- **Application source type**: The tool we use to build applications such as `Helm`, `Kustomize`, and `jsonnet`.

- **Target state**: The desired state of an application, as represented in a Git repository, which is the source of truth.

- **Live state**: The live state of that application, which means what kind of Kubernetes resources are deployed.

- **Sync status**: The status which shows that the live state matches the target state. Simply put, it's whether the application deployed in Kubernetes matches the desired states as described in the Git repository.

- **Sync**: The phase of moving an application to the target state, which happens by applying the changes in the Kubernetes cluster.

- **Sync operation status**: The status for the sync phase, whether it's failed or succeeded.

- **Refresh**: Compare the latest code in Git with the live state and check what is different.

- **Health status**: The health of the application that it is running and whether it can serve requests.

Explaining architecture

In this section, we will describe in depth the architecture of Argo CD, and we will take a deep dive into the core components of Argo CD. At the end, we will run Argo CD in a local Kubernetes cluster and run some examples to get better practical experience with it.

Architectural overview

The Argo CD core component has been implemented as a Kubernetes controller, so before diving into each component individually, we need to understand how a Kubernetes controller works.

Kubernetes controllers, in practice, observe the state of the cluster, and then they apply or request changes if they are needed. So, in practice, the controller will try to keep the current cluster state similar to the desired state. A controller observes a Kubernetes resource object – at least one – and this has a spec field that represents the desired state.

The core components of Argo CD

Argo CD is a set of various components and tools. It's time to explore in detail and learn about the responsibilities of each core component that forms the *brain* of Argo CD:

- **API server**: Similar to Kubernetes, Argo CD has an API server that exposes APIs that other systems can interact with, such as Web UI, CLI, Argo Events, and CI/CD systems. The API server is responsible for the following:

  - Managing the applications and reporting back their status

  - Triggering a set of operations to the applications

  - Management of Git repositories and Kubernetes clusters

  - Authentication and SSO support

  - RBAC policy enforcement

- **Repository server**: The repository server's main responsibility is to maintain a local cache of the Git repository holding the application manifests. Other Argo services make requests to the repository server in order to get the Kubernetes manifests, based on the following inputs:

  - The URL of the repository

  - Git revision

  - The application path

  - The template-specific settings: parameters, ksonnet environments, and `helm values.yaml`

- **Application controller**: The application controller continuously observes the live state of the application and makes a comparison with the desired state in the Git repository. Whenever there is a drift and the state are not in sync, then the controller will try to fix it and match the current state with the desired state. One of its final responsibilities is to execute any user-defined hooks for the life cycle events of the application.

A detailed overview of the architecture is represented in the following diagram:

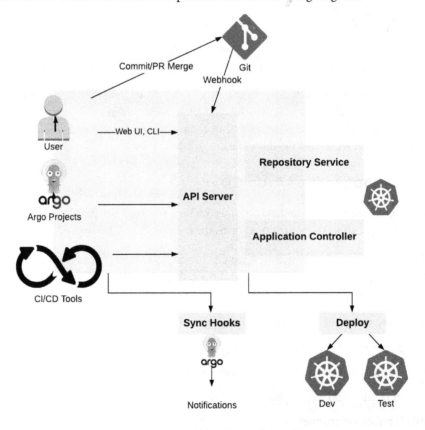

Figure 2.2 – The architecture of Argo CD

We can see in *Figure 2.2* that Argo tools and external tools interact with the API server of Argo CD directly. As described previously, Argo CD polls repositories to fetch the desired state, and by default, it polls the Git repositories every 3 minutes. If we want to avoid this delay, there are a few other options that can immediately trigger the sync phase.

Argo CD comes with a UI where we can manually initiate the sync phase, which we will explore later in this chapter.

Argo CD has a CLI to interact with the API server, as the previous diagram illustrates. For a real example, imagine that we can use the CLI from CI/CD or our local machine to sync an application with the following command:

```
argocd app sync myapp
```

The last option is to set up the Webhook configuration, as we saw in the diagram, which initiates immediately the sync phase. Argo CD supports Git Webhook notifications from GitHub, GitLab, Bitbucket, Bitbucket Server, and Gogs.

The core objects/resources of Argo CD

Argo CD applications, projects, repositories, and cluster credentials and settings can be defined declaratively using Kubernetes manifests. These are defined as CRDs. Let's take a deep dive into the core objects/resources of Argo CD:

- **Application**: Argo CD implements a CRD named `Application`, which in practice represents the instance of the application we would like to deploy in a Kubernetes cluster (environment). An example can be found as follows:

```
apiVersion: argoproj.io/v1alpha1
kind: Application
metadata:
  name: guestbook
  namespace: argocd
spec:
  project: default
  source:
    repoURL: https://github.com/argoproj/argocd-example-
apps.git
    targetRevision: HEAD
    path: guestbook
  destination:
```

```
     server: https://kubernetes.default.svc
     namespace: guestbook
```

- **AppProject**: Just like an application, there is a CRD called `AppProject`. This gives us the ability to group related applications, such as by tagging them. A real example is a separation between applications and utility services. An example can be found as follows:

```
apiVersion: argoproj.io/v1alpha1
kind: AppProject
metadata:
  name: applications
  namespace: argocd
  finalizers:
    - resources-finalizer.argocd.argoproj.io
spec:
  description: Example Project
  # Allow manifests to deploy from any Git repos
  sourceRepos:
  - '*'
  # Only permit applications to deploy to the guestbook
namespace in the same cluster
  destinations:
  - namespace: guestbook
    server: https://kubernetes.default.svc
  # Deny all cluster-scoped resources from being created,
except for Namespace
  clusterResourceWhitelist:
  - group: ''
    kind: Namespace
```

- **Repository credentials**: In real life, we need to use private repositories, and to give Argo CD the ability to access them, we need to provide some kind of access credentials. Argo CD in practice uses the Kubernetes Secrets and ConfigMaps. So, we need to create the necessary Secret Kubernetes resources with a special Kubernetes label, `rgocd.argoproj.io/secret-type: repository`. An example of a Secret can be found as follows:

```
apiVersion: v1
kind: Secret
metadata:
```

```
    name: private-repo
    namespace: argocd
    labels:
      argocd.argoproj.io/secret-type: repository
  stringData:
    url: Error! Hyperlink reference not valid.
    sshPrivateKey: |
      -----BEGIN OPENSSH PRIVATE KEY-----
      ...
      -----END OPENSSH PRIVATE KEY-----
```

- **Cluster credentials**: As with repository credentials, we need to get access to another Kubernetes cluster if Argo CD manages multiple clusters and it's not included in the ones that Argo CD already runs. The difference here is the Kubernetes label, as it's a different Secret type. `argocd.argoproj.io/secret-type: cluster`. Another example of a Secret can be found as follows:

```
apiVersion: v1
kind: Secret
metadata:
  name: mycluster-secret
  labels:
    argocd.argoproj.io/secret-type: cluster
type: Opaque
stringData:
  name: mycluster.com
  server: https://mycluster.com
  config: |
    {
      "bearerToken": "<authentication token>",
      "tlsClientConfig": {
        "insecure": false,
        "caData": "<base64 encoded certificate>"
      }
    }
```

Now that we are familiar with the important core objects and resources needed by Argo CD, it's time to run Argo CD locally and see in practice how it works.

Running Argo CD locally with Helm

Firstly, we will use Kind (`https://kind.sigs.Kubernetes.io/`) to create a new cluster locally. Create a `kind.yaml` file and use this in the configuration to create the cluster:

```
kind: Cluster
apiVersion: kind.x-Kubernetes.io/v1alpha4
nodes:
- role: control-plane
  kubeadmConfigPatches:
  - |
    kind: InitConfiguration
    nodeRegistration:
      kubeletExtraArgs:
        node-labels: "ingress-ready=true"
  extraPortMappings:
  - containerPort: 80
    hostPort: 80
    protocol: TCP
  - containerPort: 443
    hostPort: 443
    protocol: TCP
```

With the following YAML and the `node-labels` configuration, we will only allow the Ingress controller to run on a specific node(s) that matches the label selector.

In order to create the cluster, we need to run the following command and wait until the creation finishes successfully:

```
$ kind create cluster --config=kind.yaml --name=ch02
```

The output of the command should be like the following:

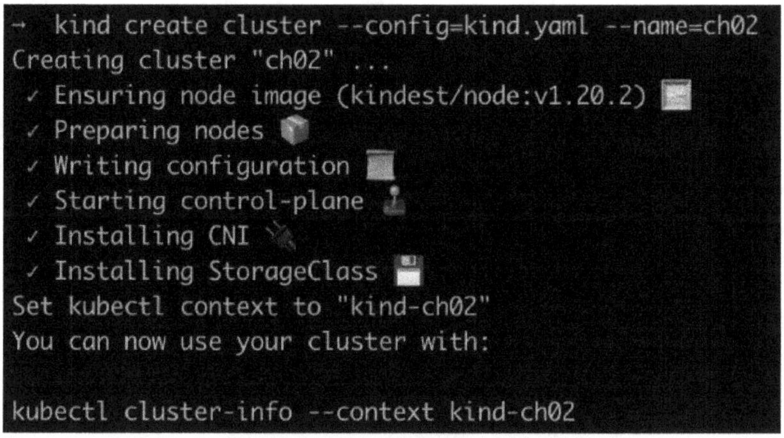

Figure 2.3 – Kind cluster status

After the creation is complete, we need to set the context of this new cluster with the following command:

```
$ kubectl config set-context kind-ch02
```

We are ready to install Argo CD using the templating tool Helm, and the chart is available here: https://github.com/argoproj/argo-helm/tree/master/charts/argo-cd.

First, we need to add the Helm repository of Argo CD to our local machine so that we can then deploy the particular chart using the following command:

```
$ helm repo add argo https://argoproj.github.io/argo-helm
```

Then, let's install Argo CD under the argocd namespace using the following command:

```
$ kubectl create namespace argocd
$ helm install ch02 --namespace argocd argo/argo-cd
```

As we didn't install any Ingress controllers, we can access the Argo CD UI, or we can interact with API through the CLI if we use port-forward the related port of the deployment:

```
$ kubectl port-forward service/ch02-argocd-server -n argocd
8080:443
```

Now that the UI is accessible, let's get the password with argocd-initial-admin-secret, which was created automatically after the Argo CD deployment:

```
$ kubectl -n argocd get secret argocd-initial-admin-secret -o
jsonpath="{.data.password}" | base64 -d
```

Running the first Argo CD application

Now, it's time to run the first application in Argo CD using the Argo CLI. First, we need to create the necessary CRD to describe an application of Argo CD and install it in the NGINX cluster:

```
apiVersion: argoproj.io/v1alpha1
kind: Application
metadata:
  name: nginx
  namespace: argocd
  finalizers:
  - resources-finalizer.argocd.argoproj.io
spec:
  syncPolicy:
    automated:
        prune: true
        selfHeal: true
    syncOptions:
      - CreateNamespace=true
  destination:
    namespace: nginx
    server: https://kubernetes.default.svc
  project: default
  source:
    chart: nginx
    repoURL: https://charts.bitnami.com/bitnami
    targetRevision: 13.2.10
```

If we check the CRD closely, we see that we are using Helm, which is one of the many ways for Argo CD to install apps. More specifically, this is the particular part in the Application CRD manifest that indicates to Argo CD that it needs to use Helm:

```
source:
  chart: nginx
  repoURL: https://charts.bitnami.com/bitnami
  targetRevision: 13.2.10
```

The next step is to apply the manifest with the `kubectl apply` command:

```
$ kubectl apply -f argo-app/
```

In the UI, hit the **SYNC** button, and if everything ran successfully, we will see a green status and the related Kubernetes resources deployed, such as the **ReplicaSet (RS)**, in the UI:

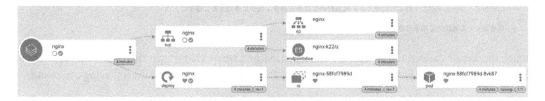

Figure 2.4 – Application synced

An alternative way to create and sync a new application without the need for a CRD is to use the Argo CLI, but this should be used in cases when we want to test something fast.

> **Important Note**
>
> The purpose of GitOps is to track everything in a Git repository and this is the source of truth when it comes to the desired state. If we just use the CLI to create Argo CD applications, then we skip this important step, and we are violating the GitOps principles.

First, we need to log in to the Argo CD server. For this example, just use the credentials from the previous **admin/admin** step:

```
$ argocd login localhost:8080
```

> **Important Note**
>
> You will get a warning from Argo CD, as by default, it doesn't use a signed certificate by a known authority. You can just type y to proceed insecurely for the sake of the experiment:
>
> **WARNING: server certificate had error: x509: certificate signed by unknown authority. Proceed insecurely (y/n)? y**

Then, type the username and password to log in, and the command output will be the following:

```
'admin:login' logged in successfully
 Context 'localhost:8080' updated
```

After the login succeeds, you are able to interact with the Argo CD API server. Now, let's create the same NGINX Argo CD application by only using the CLI:

```
$ argocd app create nginx --repo https://charts.bitnami.com/
bitnami \
  --helm-chart nginx --revision 13.2.10 \
  --dest-server https://kubernetes.default.svc \
  --dest-namespace nginx
```

Though the same as before, the application is not synced, and we need to trigger this operation explicitly. Now, we won't use the UI but a real-life example, the Argo CLI. We need to run the following command:

```
$ argocd app sync nginx
```

This command will start the sync phase of the Argo CD NGINX application, which we have set in the declarative configuration, as mentioned in *Figure 2.4*.

> **Important Note**
>
> For the sake of the example, we used the admin user. The admin user should be used only for initial configuration, and then switch to local users or configure SSO integration (https://argo-cd.readthedocs.io/en/stable/operator-manual/user-management/). More details will be described in *Chapter 3*, *Access Control*. For CI system implementations, we need to use tokens for a particular user that is responsible for the CI. An example of this is the following: $ argocd account generate-token --account ci-bot.

You made it, and this is your first experience installing Argo CD locally with Helm and your first practical deployment of Argo CD. But Argo CD is a set of tools, and the Argo CD team is trying to simplify and automate most of the manual steps we did in this section. Now, let's see how we can simplify all the steps described in this section with Argo CD Autopilot.

Running Argo CD with Argo CD Autopilot locally

If it's your first time using GitOps and Argo CD, you may have some doubts about how to structure your Git repository and manage applications across different environments. The Argo CD team created a tool called Argo CD Autopilot that can help you with onboarding to GitOps and Argo CD. Argo CD Autopilot was created to help the operator with the following:

- Creating and managing the bootstrap Argo CD application using GitOps.
- Setting up a formal structure in the GitHub repository to add new services and go through the Argo CD life cycle
- Updating and promoting applications across your different available environments

- The ability to plan for disaster recovery and bootstrap a failover cluster with all the necessary utilities and applications

- Soon to support encryption for the Secrets are used in Argo CD applications

I know you are surprised with the first bullet; it's like an *inception* where Argo CD manages itself. But the answer is, why not? Argo CD is another deployment in the cluster through manifests or Helm, so you can change its configuration accordingly using GitOps principles.

Actually, Argo CD Autopilot does a couple of magic things. The Autopilot bootstrap will push in the Git repository under a specific directory an Argo CD application manifest. This will actually manage the Argo CD installation, and you will be able to manage it with GitOps practices.

Argo CD Autopilot architecture

Again, I am a visual person, and we can understand a bit better if we check the Argo CD Autopilot architecture:

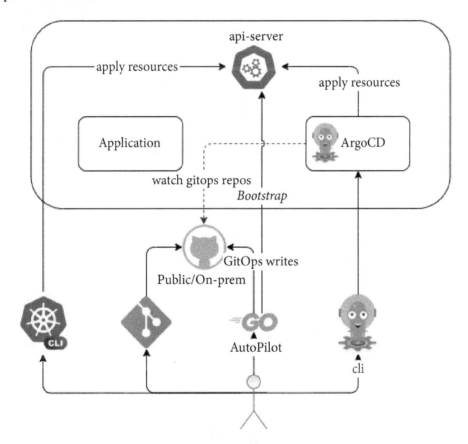

Figure 2.5 – Argo CD Autopilot architecture

During the bootstrap phase, when we deploy Argo CD, Autopilot communicates with the Kubernetes cluster, and after this, there is no need to access it anymore. The only access that will be required from now on is to the GitOps repository. When we add any new Argo CRD to the Kubernetes cluster, Autopilot will need access to the Argo CD server.

Autopilot

It's time to see a real example of how it works. First, we need to install Argo CD Autopilot, and on this page (`https://argocd-autopilot.readthedocs.io/en/stable/Installation-Guide/`), you can pick the related method according to your operating system. I prefer to use the Docker container (`https://argocd-autopilot.readthedocs.io/en/stable/Installation-Guide/#docker`) so that I can minimize the tools I install locally on my system.

Next, we need a valid Git repository where Argo CD Autopilot will push the related structure and manifests. Then, we need a valid Git token so that we can interact with the repository in terms of cloning and pushing changes. For example, if we use GitHub, we can create a token from **Developer settings** and create a personal access token (`https://docs.github.com/en/github/authenticating-to-github/keeping-your-account-and-data-secure/creating-a-personal-access-token`):

Settings / Developer settings

GitHub Apps

OAuth Apps

Personal access tokens

New personal access token

Personal access tokens function like ordinary OAuth access tokens. They can be used instead of a password for Git over HTTPS, or can be used to authenticate to the API over Basic Authentication.

Note

ARGOCD_AUTOPILOT_TOKEN

What's this token for?

Select scopes

Scopes define the access for personal tokens. Read more about OAuth scopes.

☑ repo	Full control of private repositories
☑ repo:status	Access commit status
☑ repo_deployment	Access deployment status
☑ public_repo	Access public repositories
☑ repo:invite	Access repository invitations
☑ security_events	Read and write security events

Figure 2.6 – Creating a GitHub personal access token

Next, we need to export GIT_REPO and GIT_TOKEN in the Docker container environments so that we can run Argo CD Autopilot:

```
$ export GIT_TOKEN=<your-personal-access-token>
$ export GIT_REPO=https://<your-git-repo-url>
$ argocd-autopilot repo bootstrap
```

The last command pushes the manifests, creates the necessary structure in the Git repository, and in parallel, installs Argo CD in the Kubernetes cluster for the selected context we have set for the chapter's example. You can find the structure of the repository here: https://github.com/spirosoik/ch02. After the command finishes successfully, you are ready to visit the Argo CD UI, and the output will be the following:

```
INFO argocd initialized. password: <generated password>
INFO run:
  kubectl port-forward -n argocd svc/argocd-server 8080:80
```

After running the port-forward command, use the generated password and open http://localhost:8080 to log in to the UI as admin. We will see in the UI that Argo CD manages itself already:

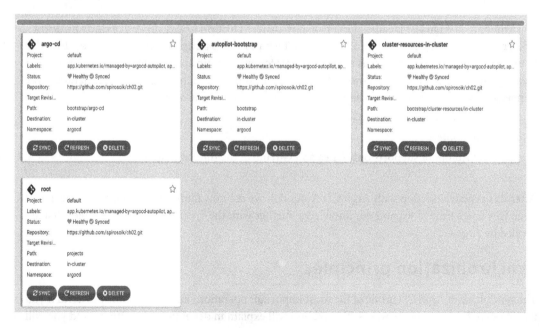

Figure 2.7 – Argo CD Autopilot inception

Now, it's time to create a testing Argo CD project and application again using Argo CD Autopilot:

```
$ argocd-autopilot project create testing
$ argocd-autopilot app create hello-world \--app github.com/
argoproj-labs/argocd-autopilot/examples/demo-app/ -p testing
--wait-timeout 2m
$ argocd-autopilot app create hello-world2 \--app github.com/
argoproj-labs/argocd-autopilot/examples/demo-app/ -p testing
--wait-timeout 2m
```

After the application is created, and after Argo CD has finished its sync cycle, our new projects will appear under the `root` application:

Figure 2.8 – Argo Autopilot app of apps pattern

This grouping in Argo CD is called the app of apps pattern (`https://argo-cd.readthedocs.io/en/stable/operator-manual/declarative-setup/#app-of-apps`). In order to understand this better, the idea is that we are able to build a main application that is responsible for creating other applications. The power of the app of apps pattern is that it gives us a declarative way to manage a group of applications and can be deployed and configured in concert.

> **Important Note**
> Argo CD Autopilot supports only Gitea and GitHub at the time of writing.

After this practical section with Argo CD Autopilot, we are now familiar with the core concepts of Argo CD, so it's time to expand our knowledge further with the synchronization principles that are applied by Argo CD.

Synchronization principles

The sync phase of Argo CD is one of the most important operations and can become powerful if you use resource hooks and sync waves. This section will explain in detail these operations, and you will see the unlimited power they can give you in a real environment.

Resource hooks

As we described in a previous section, sync is the phase for moving an application to the target state, which happens by applying the changes in the Kubernetes cluster, and this operation is executed by Argo CD in a number of steps. The phases of sync are as follows:

- Pre-sync

- Sync

- Post-sync

These are called **resource hooks**, which give us the power to run any other operation before, during, or after the sync phase:

- Use a `PreSync` hook to perform any action that needs to be done before the sync phase. A typical example is database migration as we need to run first the migrations.

- Use `Skip` to inform Argo CD to skip the application of the manifest.

- Use a `Sync` hook to orchestrate a complex deployment that requires more sophistication than Kubernetes' rolling update strategy, such as a blue/green or canary release.

- Use a `PostSync` hook to run integration and health checks after a deployment or tweet that the new release is out and any other integration with other systems.

- Use a `SyncFail` hook to run cleanup or finalizer logic if a `Sync` operation fails.

The resource hooks are applied to the particular Kubernetes manifests. For example, let's imagine that we have a schema migration for a database and we run this in a Kubernetes Kind `Job`; the resource hook will be just a Kubernetes annotation. The annotation will indicate to Argo CD when this resource should execute during the sync operation. An example of this is the following manifest:

```
apiVersion: batch/v1
kind: Job
metadata:
  generateName: schema-migrate-
  annotations:
    argocd.argoproj.io/hook: PreSync
```

The resource hooks are a nice way to reorder manifest synchronization under a logical order and based on your use case.

Sync waves

At a high level, there are three phases, pre-sync, sync, and post-sync, as described earlier. Within these three phases, we can have one or more waves that allow us to ensure that certain resources are healthy before subsequent resources are synced.

In short, you can order how Kubernetes manifests are synchronized via an annotation. Waves can assume both positive and negative values; if no wave is specified, the wave zero is assigned by default.

Configuring a wave is pretty simple; just add an annotation in the particular manifest:

```
metadata:
  annotations:
    argocd.argoproj.io/sync-wave: "5"
```

Everything works because we actually have two ways to reorder manifests:

- Resource hooks
- Sync waves

First of all, we can use both ways in parallel and combine them. When the sync operation starts and the Argo CD orders the resources with the following steps:

1. First, check the phase annotation.
2. Then the wave annotation with the lower values.
3. By the Kubernetes kind – for example, namespaces first.
4. By name.

Then, Argo CD decides which wave will be applied next, and this is where any out-of-sync or unhealthy resources will be discovered. Next, it applies this wave, and it continues in a repeated manner till all phases and waves are in sync and healthy. Let's keep in mind that if an application has resources that could be unhealthy during the first wave, then the app cannot be healthy.

We will see in practice how they work in *Chapter 5*, *Argo CD Bootstrap K8s cluster*, where we will have real-life examples of Argo CD in a production environment.

Summary

In this chapter, we got familiar with the core concepts of Argo CD, and we set up an Argo CD vocabulary so that we can have a common language in the next chapters. We checked in detail the architectural overview and how Argo CD works in detail. In parallel, we ran Argo CD in a local environment, and we deployed our first Argo CD application. Finally, we discussed two of the most powerful features of Argo CD, resource hooks and sync waves.

In the next chapter, we will start exploring how to operate Argo CD as an operator in a declarative manner and follow the best practices.

Further reading

- An Argo CD architectural overview: `https://argo-cd.readthedocs.io/en/stable/operator-manual/architecture/`

- Explaining the app of apps pattern: `https://argo-cd.readthedocs.io/en/stable/operator-manual/cluster-bootstrapping/`

- Argo CD Autopilot: `https://argocd-autopilot.readthedocs.io/en/stable/`

- SyncWaves and Hooks: `https://redhat-scholars.github.io/argocd-tutorial/argocd-tutorial/04-syncwaves-hooks.html`

Part 2: Argo CD as a Site Reliability Engineer

This part serves as an introduction to core concepts and discusses how to operate Argo CD as an SRE.

This part of the book comprises the following chapters:

3
Operating Argo CD

We will start this chapter by installing Argo CD with Kustomize using **high availability** (**HA**) manifests and go through some configuration options that we will perform while following the GitOps approach. We will make changes in the ConfigMap of a live Argo CD installation to see how we can modify different settings of Argo CD in a GitOps manner.

Then, we will look at the different Argo CD components, see what changes were introduced by the HA manifests, and what else we can do to make our installation a highly available one. While a Kubernetes cluster has a multi-control plane and worker nodes, it can still fail. Due to this, we will learn how to prepare for disaster recovery and move our installation from one cluster to another, including all the state.

Finally, we will discover what metrics are being exposed and how we can make the setup notify end users or send a custom hook to a CI/CD system when an application synchronizes successfully or not.

In this chapter, we will cover the following topics:

- Declarative configuration
- Setting up an HA installation
- Planning for disaster recovery
- Enabling observability
- Notifying the end user

Technical requirements

For this chapter, you will need access to a Kubernetes cluster. However, this time, local ones will not be enough. This is because we will be using the HA manifests, which require multiple nodes to run on so that the Pods can be spread between them. Any cluster with at least three nodes will do; the cloud provider doesn't matter. In my case, I will be using an EKS cluster from AWS, which you can set up easily with a tool such as **eksctl** (https://eksctl.io). You can think of this as a production-ready installation.

This time, we are going to install Argo CD on the cluster using Kustomize, so you will need to install it as one of your tools (https://kubectl.docs.kubernetes.io/installation/kustomize/). You will also need kubectl (https://kubernetes.io/docs/tasks/tools/#kubectl).

We are also going to make changes to some Git repositories, so Git needs to be installed (https://git-scm.com/book/en/v2/Getting-Started-Installing-Git), as well as a code editor such as **Visual Studio Code** (**VS Code**) (https://code.visualstudio.com). You will also need an account on a Git-hosted platform such as GitHub and must be familiar with working with Git commands to create commits and pull and push to remote.

The code for this chapter can be found at https://github.com/PacktPublishing/ArgoCD-in-Practice within the ch03 folder.

Declarative configuration

There are multiple ways to install Argo CD in a cluster. For one, we can directly apply its manifests, which are located at https://github.com/argoproj/argo-cd/blob/master/manifests/install.yaml – this installs the latest version. However, there are also manifests that are generated for each version, such as https://github.com/argoproj/argo-cd/blob/v2.0.0/manifests/install.yaml. Using kubectl, you will need to apply the raw manifests, so the link will be a little bit different (you will need to click on the **Raw** button after going to the preceding links):

```
kubectl apply -f https://raw.githubusercontent.com/argoproj/
argo-cd/v2.0.0/manifests/install.yaml
```

Use the official Helm chart located at https://github.com/argoproj/argo-helm/tree/master/charts/argo-cd. This option was already covered in *Chapter 2, Getting Started with Argo CD*, along with AutoPilot.

Another possibility is that, with Kustomize, similar to what's in the Argo CD repository, we have Kustomize folders (the ones where you will find a kustomization.yaml file, such as https://github.com/argoproj/argo-cd/tree/master/manifests/base). We are going to look at this option in more detail, including how to configure it and how to make it manage itself (this time, not with AutoPilot). Apart from this, we have Kustomize manifests for an HA installation. We will explore these next.

HA installation with Kustomize

For **Kustomize**, I currently have version 4.3.0 on my machine. You can find out your current version by running the following command:

```
kustomize version
```

The output of the preceding command should look similar to the following:

```
{Version:kustomize/v4.3.0
GitCommit:cd17338759ef64c14307991fd25d52259697f1fb
BuildDate:2021-08-24T19:24:28Z GoOs:darwin GoArch:amd64}
```

The code for the Kustomize installation can be found at `https://github.com/PacktPublishing/ArgoCD-in-Practice` in the `ch03/kustomize-installation` folder.

Follow these steps:

1. Create a repository where you will keep the installation configuration. This will follow the GitOps approach as every change will be done with a pull request. Because we are trying to put all our demos in the same repository for simplicity, the installation is in a folder. However, it is recommended to be in a separate repository. To use the benefits of GitOps, it is recommended not to push changes directly but to do them via pull requests so that they can be peer-reviewed.

2. In the repository, create a new folder called `resources`.

3. Inside the `resources` folder, add a new file called `namespace.yaml`. This is where we will set the namespace where Argo CD will be installed. This is the file's content:

    ```yaml
    apiVersion: v1
    kind: Namespace
    metadata:
      name: argocd
    ```

4. Directly in the root of the repository that you created earlier, add a new file called `kustomization.yaml` with the following content. As you can see, this points to the `v2.1.1` HA manifests of Argo CD (this is the latest version at the time of writing) and also references the `namespace.yaml` file we just created:

    ```yaml
    apiVersion: kustomize.config.k8s.io/v1beta1
    kind: Kustomization
    namespace: argocd
    bases:
      - github.com/argoproj/argo-cd/manifests/ha/cluster-install
    ?ref=v2.1.1
    resources:
      - resources/namespace.yaml
    ```

5. From the root of the repository, run the following command. The first part, `kustomize build .`, will generate the manifests, while the second part, `kubectl apply -f -`, will take the manifests and apply them to the cluster in a declarative manner:

```
kustomize build . | kubectl apply -f -
```

The output should start with something like this (there are many more lines; we have just provided the first seven so that you can verify this worked correctly):

```
namespace/argocd created
customresourcedefinition.apiextensions.k8s.io/
applications.argoproj.io created
customresourcedefinition.apiextensions.k8s.io/
appprojects.argoproj.io created
serviceaccount/argocd-application-controller created
serviceaccount/argocd-dex-server created
serviceaccount/argocd-redis-ha created
serviceaccount/argocd-redis-ha-haproxy created
```

6. Commit and push everything to your remote repository.

If we have a valid installation, the next step would be to check the UI, which we can do with `port-forward` (you can use another port if needed):

```
kubectl port-forward svc/argocd-server -n argocd 8085:80
```

Open your browser to `https://localhost:8085`; you should see the login page, which we already discussed in *Chapter 2, Getting Started with Argo CD*, and how to connect (and change the admin password). You will probably get a certificate warning because Argo CD is using self-signed certificates, but at this time, there should be no risk in visiting the website. For a production installation, you would also need a `LoadBalancer` service to be created with a certificate attached so that you can offload TLS to it.

In the next section, you will learn how to turn Argo CD into an application that can be managed via Argo CD itself, allowing for easy and declarative configuration updates.

Argo CD self-management

Argo CD can manage itself in the same way it manages other applications from the cluster. This is similar to what AutoPilot (`https://argocd-autopilot.readthedocs.io/en/stable/`) does under the hood. In this section, we will create an Argo CD application that points to the folder where we have our **Kustomize** manifests. This way, Argo CD will start monitoring that repository and folder for changes. Any new commits we make to the folder will be applied automatically.

First, we must create a file called `argocd-app.yaml` in our repository with the following content. An Argo CD application is made up of three parts: `destination`, which is where the manifests are applied, `project`, which we use to create specific restrictions (for example, this application should only deploy resources to a cluster and a specific namespace), and the `source` repository, which where it takes the manifests, including the `branch` and `repo` folder:

```yaml
apiVersion: argoproj.io/v1alpha1
kind: Application
metadata:
 name: argocd
spec:
 destination:
    namespace: argocd
    server: https://kubernetes.default.svc
 project: default
 source:
    path: ch03/kustomize-installation
    repoURL: https://github.com/PacktPublishing/ArgoCD-in-
Practice.git
    targetRevision: main
```

Now, we are going to apply this application with `kubectl`:

```
kubectl apply -f argocd-app.yaml -n argocd
```

The output should be similar to the following:

```
application.argoproj.io/argocd created
```

By creating this application, we are telling Argo CD that, in the `https://github.com/PacktPublishing/ArgoCD-in-Practice.git` repository, in the `ch03/kustomize-installation` folder, some manifests should be applied. These are the manifests for the Argo CD installation. In terms of the control loop, the state that we desire is the state from the cluster, so after the observe state phase, there should be no actions to take. The main thing is that from now on, Argo CD monitors the repository every 3 minutes (by default) and checks for new commits. If it finds any, it will recalculate the manifests and try to apply them to the cluster. This is the mechanism through which Argo CD can manage itself.

If you have stopped `port-forward`, which we enabled earlier, you can reenable it with the following command:

```
kubectl port-forward svc/argocd-server -n argocd 8085:80
```

Then, if you go to `https://localhost:8085`, you will see your first application that has been synchronized by Argo CD, which is Argo CD itself:

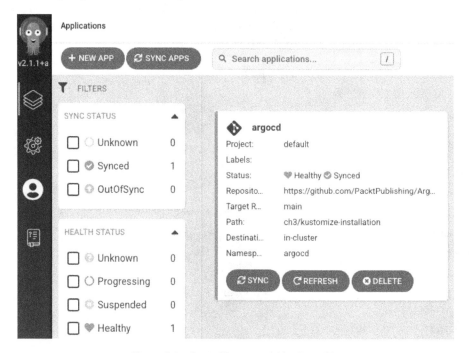

Figure 3.1 – Argo CD managed by Argo CD

Now, let's look at some simple configuration updates we can apply automatically to Argo CD by creating a commit and pushing it to remote.

Configuration updates

Since version 2.1 of Argo CD, we have a new setting in the main ConfigMap that allows us to modify the default interval that's used to check for new updates on Git repositories. Every 180 seconds, it checks if new commits have been pushed. We can modify this using the `timeout.reconciliation` parameter. Before this parameter was introduced, we had to change `StatefulSet` of the application controller to set a different value with the `–app-resync` flag (which has been deprecated since version 2.1).

To update this reconciliation timeout, we will create a new folder called `patches`, at the same level as the `resource` folder. Inside it, we will create a new file called `argocd-cm.yaml`. We are not replacing the entire ConfigMap; instead, we are applying some patches to it – in this case, just the `timeout.reconciliation` field. The content of this new file will be as follows:

```
apiVersion: v1
kind: ConfigMap
```

```
metadata:
  name: argocd-cm
data:
  timeout.reconciliation: 300s
```

We also need to update the kustomization.yaml file by adding a reference to the change that we made for our ConfigMap. The file should now look like this:

```
apiVersion: kustomize.config.k8s.io/v1beta1
kind: Kustomization
namespace: argocd
bases:
  - github.com/argoproj/argo-cd/manifests/cluster-
install?ref=v2.1.1
resources:
  - resources/namespace.yaml
patchesStrategicMerge:
  - patches/argocd-cm.yaml
```

Next, we need to create a commit that contains the new argocd-cm.yaml file and the change we made to kustomization.yaml and then push it to the remote repository. Argo CD is checking for new commits every 180 seconds, so it will identify the change and apply it. After doing so, it will check for new commits every 300 seconds.

The documentation says that it is not enough to update this setting – we also need to manually restart the argocd-repo-server deployment for the new configuration to be loaded. This can be done using the following command:

```
kubectl rollout restart -n argocd deployment argocd-repo-server
```

The output should be similar to the following:

```
deployment.apps/argocd-repo-server restarted
```

Having Argo CD manage itself is useful as we can make changes to its running installation, from small configuration updates, as we just saw, to upgrading the version. And all these will be applied automatically by Argo CD, making it become an application like any other managed by Argo CD. The normal GitOps flow should include creating a pull request with your changes so that they can be reviewed by your peers. In our case, for simplicity, we are pushing directly to the remote default branch, so they will be applied right away. Next, we will discover how an HA installation can be achieved by looking at all the different Argo CD components and the changes we will apply to them.

Setting up an HA installation

Since we have already used the HA option with Kustomize, let's see what components are installed, how they handle the HA part, and if there is anything else we can do:

- **API server**: This handles all the exterior interaction, so if you are using the CLI or the UI or creating a client, you will communicate with the API. The HA manifests already set two instances for this pod.

- **Repository server**: This is responsible for creating the final manifests to apply to the cluster; manifests generation is complicated because of all the templating supported by Argo CD, such as Helm 2 or 3, Kustomize, and Jsonnet. The HA manifests come with two replicas.

- **Application controller**: This is where the work is initiated, where the control loop is implemented, and application sync takes place. Initially, you were only able to have one instance, but now, you can have one instance per cluster shard. The HA manifests use one instance of the controller.

- **Redis cache**: Manifest generation is an expensive operation and Argo CD tries to save the manifests in a Redis instance; there is no problem if the cache fails as it can be recalculated, but do expect a performance penalty. Here, we probably have the biggest changes between the normal and HA manifests. In HA mode, we get an additional deployment of HAProxy and three replicas for Redis – that is, one master and two slaves.

- **Dex server**: This is responsible for user authentication when you use external identity providers such as **Security Assertion Markup Language (SAML)**, **OpenID Connect (OIDC)**, or **Lightweight Directory Access Protocol (LDAP)**. It is optional, but if you want to, for example, connect your GitHub or Google account to Argo CD, you will need this component.

Let's discuss each in terms of HA.

API server

The API server is the entry point for all our requests, regardless of whether they are coming from the UI, CLI, or a custom client, such as curl. It doesn't have any state, so we can scale it up or down, depending on the load. This also means we can keep an HA installation just by changing the number of replicas for its deployment. By using the HA option, we got two replicas, but let's see how we can update this number to three and what other changes need to be made.

Besides the replicas, optionally, we could update the ARGOCD_API_SERVER_REPLICAS environment variable so that it has the same number of replicas we are using. This is used when calculating a limit for brute-force password attacks. Let's say that if, for one instance, 30 concurrent login requests would be a lot to trigger a different response from the server, for three instances, the load will be spread, so you will only get 10. This means that the more instances we have, the lower we need to go with the limit. Now, if you don't update the variable, the application will still work, but if you do update it, then it will be safer.

To have three replicas of the `argocd-server` deployment, we need to do the following. Create a new file called `argocd-server-deployment.yaml` in the `patches` folder. Its content should be as follows:

```yaml
apiVersion: apps/v1
kind: Deployment
metadata:
  name: argocd-server
spec:
  replicas: 3
  template:
    spec:
      containers:
        - name: argocd-server
          env:
            - name: ARGOCD_API_SERVER_REPLICAS
              value: '3'
```

Then, we need to add the entry to the `kustomization.yaml` file so that our change will be taken into consideration. I am not going to show the whole file here, just the section starting from `patchesStrategicMerge`:

```yaml
patchesStrategicMerge:
  - patches/argocd-cm.yaml
  - patches/argocd-server-deployment.yaml
```

Now, as usual, we must create a `git commit` with these two files and then push it to remote so that Argo CD can see the new revision and apply the changes to the installation.

Repository server

The **repository** (**repo**) server is an important component that generates the resources to apply to a cluster. Usually, in our GitOps repos, we don't use simple manifests; instead, we use templating engines such as Helm, Jsonnet, and Kustomize. This component is transforming those templates into manifests that are ready to be applied with the `kubectl apply` command. It goes and fetches the content of the Git repo, based on which it knows whether to use Helm, Kustomize, or something else (for example, if it finds a file called `Chart.yaml`, it knows it is a Helm chart). After it finds out what the templating engine is, it runs commands such as `helm template` and `kustomize build` to generate the final manifests. For Helm, it may need to do a `helm dep update` beforehand to fetch any external dependencies.

A lot is going on in this repo server application, which means that if we run multiple instances of it, we will be able to generate more manifests in parallel. It also makes sense to provide enough resources so that these containers don't get killed because of out-of-memory errors or throttled on the CPU. If you have thousands of applications that have been deployed with Argo CD, you can easily run more than 10 instances of the repo server and assign something such as 4 to 5 CPUs and 8 to 10 GB of memory to each instance. From the HA manifests, we have two instances already, but we are going to modify it so that it has three. We will not place any resource requests or limits because we are using local clusters, but for real ones, this is highly recommended.

> **Note – Repo server performance**
>
> My experience with repo servers is highly impacted by the usage of Helm 2. When we migrated most of our charts to Helm 3, we ran some tests and realized that the move significantly reduces the manifest generation time (at least in some situations, which still occur pretty often in our setup). So, you should take the preceding container resource recommendations with a pinch of salt and do your own calculations.

Another important parameter that we may need to modify is the timeout for the templating engine. Argo CD forks the Helm or Kustomize commands and sets a 90-second timeout for these operations. Sometimes, that may not be enough, such as when Kustomize is using big remote bases or when Helm needs to template big charts such as `kube-prometheus-stack` (`https://github.com/prometheus-community/helm-charts/tree/main/charts/kube-prometheus-stack` – it used to be known as `prometheus-operator`) or `istio` (`https://istio.io/latest/docs/setup/install/helm/`). So, you can run some tests and see what would work for you. If you want to increase the timeout, you can use the `ARGOCD_EXEC_TIMEOUT` environment variable.

Create a new file called `argocd-repo-server-deployment.yaml` in the `patches` folder with the following content, where 3 replicas have been set for the repo server and 3 minutes have been set for the templating timeout:

```
apiVersion: apps/v1
kind: Deployment
metadata:
 name: argocd-repo-server
spec:
 replicas: 3
 template:
   spec:
     containers:
       - name: argocd-repo-server
```

```
      env:
        - name: "ARGOCD_EXEC_TIMEOUT"
          value: "3m"
```

We also need to update the `kustomization.yaml` file so that it includes a reference to the new file we just created in the `patches` folder (I am adding the `patchesStrategicMerge` section here, which includes the change and not the entire content of the file):

```
patchesStrategicMerge:
  - patches/argocd-cm.yaml
  - patches/argocd-server-deployment.yaml
  - patches/argocd-repo-server-deployment.yaml
```

Create the Git commit and push it to remote, so that Argo CD can then update the repo server deployment.

Application controller

Initially, this was the component that couldn't have more than one replica because there was a control loop that was the initiator of all synchronizations. Thus, having more than one replica would introduce the possibility of two or more synchronizations starting for the same application at the same time. But starting with version 1.8, you can have more than one replica, with each instance taking care of a part of the clusters registered in Argo CD.

For example, if you have nine clusters where Argo CD is installing applications and you start three application controllers, then each controller will take care of three of those clusters. Scaling them is not necessary for an HA installation, but it does help in this direction because the failure of one controller only affects part of the clusters, not all of them. It also helps the overall performance of Argo CD by splitting the load into multiple instances.

To tell the Argo CD application controller how many shards (or instances) it can have, you can use the ARGOCD_CONTROLLER_REPLICAS environment variable from its `StatefulSet`. Let's look at what an overlay for the Kustomize installation set to 3 replicas (which would mean three shards) for the controller would look like. Create a file called `argocd-application-controller-statefulset.yaml` under the `patches` folder with the following content:

```
apiVersion: apps/v1
kind: StatefulSet
metadata:
  name: argocd-application-controller
spec:
  replicas: 3
  template:
```

```
spec:
  containers:
    - name: argocd-application-controller
      env:
        - name: ARGOCD_CONTROLLER_REPLICAS
          value: "3"
```

Now, update the `patchesStrategicMerge` section of the `kustomization.yaml` file, like this:

```
patchesStrategicMerge:
  - patches/argocd-cm.yaml
  - patches/argocd-server-deployment.yaml
  - patches/argocd-repo-server-deployment.yaml
  - patches/argocd-application-controller-statefulset.yaml
```

Create the commit, push to remote, and Argo CD will take care of the rest.

There might be scenarios when there is only one destination cluster, meaning all your applications, regardless of whether they are dev, test, QA, or prod, will be installed on one cluster only. In this case, it doesn't make sense to have more than one instance of application controllers, but you should allocate a large amount of CPU and memory for the container. You should check the `--operation-processors`, `--status-processors`, and `--kubectl-parallelism-limit` flags in the official documentation (`https://argo-cd.readthedocs.io/en/stable/operator-manual/high_availability/#argocd-application-controller`) and set higher values for them to allow your instance to handle more applications.

> **Note – Replicas in environment variables**
>
> This pattern is used in at least two places: the API server and the application controller. Here, the number of replicas is injected into the container with an environment variable. This way, it is a lot simpler compared to calling the Kubernetes API from each instance to find out the number. Even with the additional overhead for the developer to make sure they update both places, it is still worth doing it like this.

Redis cache

Redis is used by Argo CD as a throwaway cache, meaning the system still works if Redis is not responding, gives some error, or isn't installed at all, though this might affect its performance. So, this is an optional component, but a highly recommended one.

This is because the manifests that were generated from a Git repository will be kept in the Redis cache, so if Redis is missing, they will have to be recreated at every synchronization request. The cache is only dropped if there are new commits to the Git repo (think of the commit's SHA as the key). If the cache is lost, everything needs to be recreated, which means the application still works but with a performance hit.

The HA installation comes with a `StatefulSet` with three replicas for Redis – one master and two slaves. It also comes with an HAProxy deployment that sits in front of Redis. If the Redis master fails for some reason and one of the slaves is promoted as the new master, HAProxy will make this transparent to the client applications.

In terms of HA, the Redis installation has a very good setup, so there is no need to make any changes.

> **Note – Redis manifests**
>
> The Redis manifests that are used by Argo CD are generated from a Helm chart located at `https://github.com/DandyDeveloper/charts/tree/master/charts/redis-ha`. When we use Kustomize or the simple manifest options, the Helm chart is already templated and transformed into simple resources. The Helm installation has a `redis-ha` chart declared as a dependency, so it is consumed directly.

Dex server

Dex (`https://github.com/dexidp/dex`) is used to delegate authentication when an external system is involved, such as when using something such as OIDC. Because dex keeps an in-memory cache, we can't use more than one instance; otherwise, we risk inconsistencies. What we risk if it goes down is not being able to log in using the external system. This is not critical as Argo CD continues to do all the necessary reconciliation, so it should still have connectivity to the Git repos and the destination Kubernetes clusters, which means its work will not stop. The login downtime should be temporary because by being installed as a one-replica deployment, the controller will restart the instance (sometimes, it will do this with our help when node issues are involved).

If you only use local users (we will learn more about them in *Chapter 4, Access Control*) or the admin special user, then you can disable the dex installation by setting the number of replicas to zero.

HA is one of the best practices for reducing the risk of interruptions for your services. Even if your Argo CD instance is down for a small period, you don't want this to happen while doing any kind of production deployment or rollback. So, it becomes critical that we eliminate the single points of failure by building redundancy and resilience into Argo CD components. Luckily, we get HA manifests out of the box. Once we understand how each component is modified to become highly available, we can take more steps toward improved service, from using even more replicas than the default to splitting the Kubernetes clusters where we are deploying our applications to more application controllers. Next, we are going to look at disaster recovery, which is about getting the system back to a working state after it becomes inoperative. This can help us put things back in order where HA was not enough.

Planning for disaster recovery

Argo CD doesn't use any database directly (Redis is used as a cache), so it looks like it doesn't have any state. Earlier, we saw how we can have a high availability installation, done mostly by increasing the number of replicas for each deployment. But we also have application definitions (such as the Git source and destination cluster) and details on how to access a Kubernetes cluster or how to connect to a private Git repo or a private Helm one. These things, which make up the state of Argo CD, are kept in Kubernetes resources – either native ones, such as secrets for connection details, or custom ones for applications and application constraints.

A disaster may occur due to human intervention, such as the Kubernetes cluster or the Argo CD namespace being deleted, or maybe some cloud provider issues. We may also have scenarios where we want to move the Argo CD installation from one cluster to another. For example, maybe the current cluster was created with a technology that we don't want to support anymore, such as **kubeadm** (https://kubernetes.io/docs/setup/production-environment/tools/kubeadm/), and now we want to move to a cloud provider-managed one.

A question may have popped up in your head: "*but I thought this is GitOps, so everything is saved in Git repos, which means that it is easy to recreate?*" First, not quite everything gets saved into Git repos. For example, when registering a new cluster in Argo CD, we do so imperatively by running a command so that such details are not in Git (which is OK for security reasons). Second, recreating everything from the GitOps repos may take a lot of time – there might be thousands of applications, hundreds of clusters, and tens of thousands of Git repos. A better option might be to restore all the previous resources from a backup instead of recreating all of them from scratch; that would be much faster.

Installing the CLI

Argo CD provides a utility part of the main CLI (the `argocd admin` subcommand) that can be used to create the backup (export all related data) to a YAML file or import data from an existing one. The CLI can be found in the main Docker image, or it can be installed separately.

To install the v2.1.1 version of the CLI, run the following commands (this is the macOS version; for other options, please check the official page: https://argo-cd.readthedocs.io/en/stable/cli_installation/):

```
curl -sSL -o /usr/local/bin/argocd https://github.com/argoproj/
argo-cd/releases/download/v2.1.1/argocd-darwin-amd64
chmod +x /usr/local/bin/argocd
argocd version --client
```

If everything went OK, then the output of the preceding commands should show the Argo CD client version, like so:

```
argocd: v2.1.1+aab9542
  BuildDate: 2021-08-25T15:14:05Z
  GitCommit: aab9542f8b3354f0940945c4295b67622f0af296
  GitTreeState: clean
  GoVersion: go1.16.5
  Compiler: gc
  Platform: darwin/amd64
```

Now that we have installed the CLI, we can use it to create the backup.

Creating the backup

Now, we can connect to our cluster and create the backup. You should be connected to the cluster where you have the Argo CD installation (your kube context for pointing to the cluster). Run the following command, which will create a file with a custom name based on the current date and time (this way, you can run it daily or even more frequently):

```
argocd admin export -n argocd > backup-$(date +"%Y-%m-
%d_%H:%M").yml
```

Even if we had a simple installation with just one application (Argo CD itself), you can see that the backup is quite a big file (it should contain almost 1,000 lines). This is also because the Argo CD application is a pretty big one (we deployed many resources) and it keeps a history of everything it has synchronized. You will find the backup file that I generated for the HA installation in this book's Git repository (https://github.com/PacktPublishing/ArgoCD-in-Practice) in the ch03/disaster-recovery folder.

Next, we should take this backup file and save it on a cloud storage system (such as AWS S3, Azure Blob, or Google Cloud Storage), encrypt it, and have access policies around it. This is because, for a real installation, a lot of sensitive information will be stored there, including access to your production Kubernetes clusters.

Restoring on a different cluster

To restore a backup, you need to install Argo CD in the destination cluster. This is because, in the backup, we have its configuration, as well as all the ConfigMaps and Secrets, so everything we changed for the initial installation should be present. However, the backup doesn't store the actual Deployments or StatefulSets. This means they need to be installed before the backup is restored. The same goes for CustomResourceDefinitions – we will have all the instances of Application and AppProject, but we won't have the definitions of those custom resources.

So, in the new cluster, perform the same installation we did previously in the *HA installation with Kustomize* section. After that, run the following command (you will need to change the filename so that it matches yours):

```
argocd admin import - < backup-2021-09-15_18:16.yml
```

Now, you should have a new installation with all the states (applications, clusters, and Git repos) you had when you created the backup. The only difference is that now, there is nothing in the Redis cache, so Argo CD will need to start recalculating all the manifests from the Git repos, which may affect the system's performance for the first couple of minutes. After that, everything should be business as usual.

In this section, we saw that Argo CD tools are easy to automate, from creating regular backups to restoring them on newly created clusters. It is important to have a backup strategy and to do restoration exercises from time to time. We should be prepared for when a disaster occurs and have the necessary runbooks so that we can execute them with the same results, regardless of whether it is 2:00 A.M. or 2:00 P.M. Disasters are rare, so we will come across a lot of situations in our day-to-day operations. This could be an increased number of applications to sync or a specific version of a YAML templating tool leading to timeouts or even an unresponsive system. For this, we need to have a good strategy for observability. We will explore this in the next section.

Enabling observability

Observability is important because it can provide answers about the system's health, performance, and behavior. When working in a big setup with thousands of applications with tens of teams deploying their monoliths and microservices to Kubernetes, there's a big chance that things won't always go as smoothly as you would expect. There is always some wrong setting, an older version that wasn't supposed to be used, an immutable field trying to be updated, many applications needing to be synchronized at the same time, a team trying to use a private repo without setting its SSH key, or big applications that can cause timeouts.

Luckily, Argo CD exposes many metrics that allow us to understand the system, if it is underutilized or overutilized, and what to do about it. It also gives us ways to directly alert the development teams that are responsible for specific applications when something goes wrong, such as a failed synchronization. The alerts that we will create can be split into two directions – one for the team responsible for operating Argo CD and one for the teams taking care of the microservices.

In this section, we will learn how to monitor Argo CD with Prometheus, which has become the default choice for monitoring dynamic environments such as microservices running in containers on Kubernetes. Prometheus, with its focus on reliability, is one of the best tools to find out the current state of your system and to easily identify possible issues.

Monitoring with Prometheus

Like Kubernetes became a standard for container orchestration, Prometheus became a standard for monitoring. It was the second project to enter **Cloud Native Computing Foundation** (**CNCF**), with Kubernetes being the first. In the cloud-native world, we have an operator for running Prometheus in Kubernetes (like Argo CD is an operator for GitOps) called **Prometheus Operator** (`https://prometheus-operator.dev`). Argo CD components expose metrics in Prometheus format, making it easy to install Prometheus Operator in the cluster and start scraping those endpoints. There is a Helm chart you can use to install it (usually, this is done in a separate namespace called **monitoring**): `https://github.com/prometheus-community/helm-charts/tree/main/charts/kube-prometheus-stack`.

After the installation, we will need to tell Prometheus where it can find the endpoints that expose the metrics. To do that, we can use the custom `ServiceMonitor` resource (`https://prometheus-operator.dev/docs/operator/design/#servicemonitor`). Three services should be scraped – one for the application controllers, one for the API servers, and one for the repo servers – thus covering all the Argo CD components. You can find the `ServiceMonitor` resources in the official documentation at `https://argo-cd.readthedocs.io/en/stable/operator-manual/metrics/#prometheus-operator`. We also saved a copy of them in our Git repo (`https://github.com/PacktPublishing/ArgoCD-in-Practice`) in the `ch03/servicemonitor` folder.

You can apply the files by putting them in a Git repo, inside a folder, and then creating an application that points to it so that you can apply them using GitOps.

After we have the `ServiceMonitor` resources in place and the scraping process has begun, there is a Grafana dashboard (`https://grafana.com/grafana/dashboards`), which is offered as an example in the official documentation at `https://argo-cd.readthedocs.io/en/stable/operator-manual/metrics/#dashboards`, that you can use. Please follow the official documentation on how to import a dashboard to see how you can add this to your own Prometheus Operator installation: `https://grafana.com/docs/grafana/latest/dashboards/export-import/#import-dashboard`. We will cover monitoring from two different perspectives – one for the team taking care of Argo CD, which we are calling the operating team, and the second for the team building applications, which we are calling the microservices team.

Metrics for the operating team

To perform synchronizations, Argo CD uses the repo servers and the controllers. These are the most important pieces that we need to monitor and taking good care of them will allow us to have a good, performant system. Various metrics can help us understand their behavior. So, let's explore a few of them.

OOMKilled

Over time, we realized that the most valuable metric for these two components was not something exposed by Argo CD but the **out-of-memory** (**OOM**) kills that were performed by the node OS on the containers trying to use too many resources. This is a good indicator of whether the container resources that you have set are not enough or if you are using a parameter that sets a value that's too big for parallelism. The Argo CD documentation provides a good explanation of when OOM can occur and what parameters to use to reduce parallelism: `https://argo-cd.readthedocs.io/en/stable/operator-manual/high_availability/`. In terms of the repo server, too many manifests are generated at the same time, while for the application controller, too many manifests are being applied at the same time.

You can use the following query to get alerts about these events. It checks if the containers have been restarted in the last 5 minutes and whether the last terminated reason was `OOMKilled` (this query comes from this old and valuable Kubernetes thread: `https://github.com/kubernetes/kubernetes/issues/69676#issuecomment-455533596`):

```
sum by (pod, container, namespace) (kube_pod_container_
status_last_terminated_reason{reason="OOMKilled"}) *
on (pod,container) group_left sum by (pod, container)
(changes(kube_pod_container_status_restarts_total[5m])) > 0
```

If you receive such alerts once or twice a week, that might not be that bad, and you can wait a few more weeks to see what happens. If they occur a couple of times per day, either for the repo servers or the controllers, you should take action. Here are a few things you can do:

- Increase the number of replicas for `Deployment/StatefulSet` so that when an application synchronization needs to occur, the load will be spread to more instances.

- Set more resources – both CPU and memory – for the containers.

- Reduce the value of the `--parallelismlimit` parameter for the repo servers and the value of the `--kubectl-parallelism-limit` parameter for the controllers.

OOM is related to how much work the controller needs to do to reconcile the application's state. This means that if you don't have deployments for a few days, this probably won't happen, while if you start synchronizing many applications at once, you may start receiving OOM alerts. If this is the case, then we should see a correlation with the load metrics we have defined in the system. We will look at those next.

System load metrics

There are a few metrics that can reveal the load of the system. Here, we will look at one that's relevant for repo servers and one that's relevant for application controllers.

The repo servers are tasked with fetching the Git repos' content and then creating the manifests based on the templating engine that was used. After they create the final manifests, their work is continued by the application controllers. We have seen that too many manifests using `apply` at the same time can lead to OOM issues, but what happens when we have a lot of requests for fetching the content of Git repos? In this case, there is a metric called `argocd_repo_pending_request_total` that goes up and down (in Prometheus, we would call it a gauge), depending on how many requests are pending on the repo server instances. Of course, the number should be as close to zero as possible, showing that the current number of repo instances can handle the load. However, this is not a problem if it goes up for a short period. Problems can occur when there is a big value for a long period, so that's what you should look out for.

> **Note – Scaling repo servers with HPA**
>
> If you are already thinking about scaling repo servers with HPA based on this gauge, please join the discussion on this thread as it turns out it is not that easy: `https://github.com/argoproj/argo-cd/issues/2559`.

On the application controller side, there is another important metric for showing the system load – that is, `argocd_kubectl_exec_pending`. This shows how many `apply` and `auth` commands are pending to be executed on the destination clusters. The maximum number can be equal to the `--kubectl-parallelism-limit` flag because that's how many parallel threads can start commands on the destination clusters. It is not a problem if it reaches the maximum for a short period, but problems such as synchronization taking a lot of time can occur when the value remains big for a longer period of time.

Metrics for the microservices teams

If you are trying to apply the idea of platform teams that create self-service platforms for development teams, then you should allow the development teams to monitor, get alerts, and take action when something goes wrong with their live deployments. One way to do this is to allow them to set up alerts for the Argo CD applications that are used to take their microservices to production. Two metrics can provide value for development teams. One can be used for the synchronization status, especially if there was a failure during the synchronization process. The other can be used for the application's health status, especially the `Degraded` status, which means something is not functioning as expected.

Application synchronization status

The synchronization status is useful to alert on so that the team doesn't need to pay attention to the UI or run regular commands via the CLI to find out the status of the new version deployment. This is especially true when this is done several times per week, not to mention if you do it more frequently. The teams can set up alerts for the applications that they manage so that if they fail to synchronize the new version of their Docker image or some other change they made to the manifests, then they will receive an alert. The `argocd_app_sync_total` metric can be used for this.

The following query can be used to alert you about any applications that have changed their synchronization statuses to `Failed` in the last 5 minutes. The query only looks for applications from the `argocd` namespace and that start with `accounting` (which would be of interest to an accounting team):

```
sum by (name) (changes(argocd_app_sync_total{phase="Failed",
exported_namespace="argocd", name=~"accounting.*"}[5m]))>0
```

If there are no issues, we shouldn't get any results. However, if we do get any applications with a `Failed` synchronization status, we should start investigating the reasons.

Application health status

The health status is different than the synchronization status because it can be modified independently of a synchronization. We are usually looking for a `Degraded` status, which happens when something is not functioning correctly, such as if you asked for three replicas in your `StatefulSet`, but only two are up and running while the third one is still initializing after a long period, or it was terminated and now it is not being scheduled and remains `Pending`. Such scenarios can occur at any time while the application is running in production and it is not directly related to a synchronization event. The metric for tracking this is `argocd_app_info`. You can use the following query to track the `Degraded` status of an Argo CD application from the `argocd` namespace if its name starts `prod` but it doesn't end with `app` (this can be useful when the intermediary applications that were created by the app-of-apps pattern use the app suffix: `https://argo-cd.readthedocs.io/en/stable/operator-manual/cluster-bootstrapping/#app-of-apps-pattern`):

```
argocd_app_info{health_status="Degraded",exported_
namespace="argocd",name=~"prod.*",name!~".*app"}
```

Getting applications with `Degraded` statuses in the results is a clear sign that some problems in the cluster are preventing your application from functioning properly, so it needs to be checked.

Next, we will learn how to notify the user about events that are happening in Argo CD, such as whether an application was deployed successfully. This can be achieved with different tools. We are going to look at the ones that are Argo CD-specific, such as the Argo CD Notifications project and the custom webhooks built into Argo CD.

Notifying the end user

To synchronize applications, Argo CD can work in two different ways. First, it can work manually so that a new commit to the GitOps repo will not have any direct effect unless you trigger the synchronization manually via the CLI, by using the UI, or by using an API call. The second mode, which I think is the most used one, is that after a push to the repo, Argo CD will start to automatically reconcile the cluster state so that it matches the one we declared.

The developers that performed the state changes are interested in the outcome of the reconciliation – they want to know if their microservices are up and running correctly or if they have some problems with the new configuration or the new container image.

Earlier, we learned how to monitor the synchronization process using Prometheus and the metrics Argo CD exposes for application health and synchronization status. However, there is another way we can notify development teams that their microservices have some failures or when everything went perfectly: the Argo CD Notifications project. This was built especially with Argo CD in mind and can provide more useful details to its users. You can learn more about the Argo CD Notifications project at `https://github.com/argoproj-labs/argocd-notifications`.

Installing Argo CD Notifications

Like Argo CD, the Notifications project can be installed in three different ways: via a Helm chart (`https://github.com/argoproj/argo-helm/tree/master/charts/argocd-notifications`), using simple manifests (`https://github.com/argoproj-labs/argocd-notifications/tree/master/manifests`), or via Kustomize. We will go with the Kustomize option and install it using GitOps mode. All the code for the notifications we will build can be found at `https://github.com/PacktPublishing/ArgoCD-in-Practice.git` in the `ch03/notifications` folder.

In the same repo that you used to install Argo CD, create a new folder called `notifications`. Inside that folder, create a file called `kustomization.yaml` and add the following content. As you can see, we kept the same `argocd` namespace; there is no reason to create another one as this is not a standalone application. However, it needs an instance of Argo CD to work with (in other words, we won't have an instance of Argo CD Notifications if Argo CD isn't installed):

```
apiVersion: kustomize.config.k8s.io/v1beta1
kind: Kustomization
namespace: argocd
bases:
  - github.com/argoproj-labs/argocd-notifications/manifests/
controller?ref=v1.1.1
```

You should commit the file to your repo and then push to remote so that we can create the application file. Name this `argocd-notifications-app.yaml` and place it in the top folder this time (it should be at the same level as the `argocd-app.yaml` file that we created earlier in this chapter when we made Argo CD self-managed). Here is the content of the file (just make sure that you replace `path` and `repoURL` with your own values):

```
apiVersion: argoproj.io/v1alpha1
kind: Application
```

```
metadata:
  name: argocd-notifications
spec:
  destination:
    namespace: argocd
    server: https://kubernetes.default.svc
  project: default
  source:
    path: ch03/notifications
    repoURL: https://github.com/PacktPublishing/ArgoCD-in-
Practice.git
    targetRevision: main
  syncPolicy:
    automated: {}
```

Now that we've created the file, we need to apply it with the following command (don't forget to commit and push it to your repo for future reference):

```
kubectl apply -n argocd -f argocd-notifications-app.yaml
```

After applying it, you should have the Argo CD Notifications application installed by Argo CD in your cluster. In the UI, it should look like this:

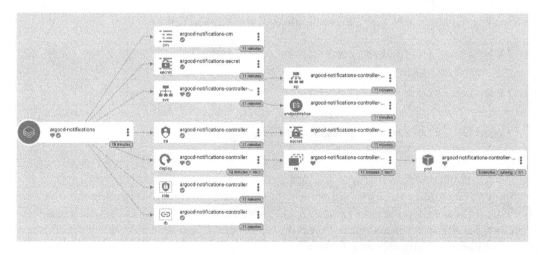

Figure 3.2 – The Argo CD Notifications application in the Argo CD UI

Next, we'll learn how to start GitLab pipelines from Argo CD Notifications. This can be done to update the deployment status of our application in various tracking systems and can be seen as a way to close the GitOps reconciliation loop.

Starting a pipeline

The following is a good article on the Argo blog that explains how to set the proper configurations to send notifications via emails: `https://blog.argoproj.io/notifications-for-argo-bb7338231604`. In the official documentation, there are more examples of notification services such as Slack (`https://argocd-notifications.readthedocs.io/en/latest/services/slack/`), Microsoft Teams (`https://argocd-notifications.readthedocs.io/en/latest/services/teams/`), Telegram (`https://argocd-notifications.readthedocs.io/en/latest/services/telegram/`), and a few others. There are also good explanations on how to use webhooks to set statuses on GitHub commits (`https://argocd-notifications.readthedocs.io/en/latest/services/webhook/#set-github-commit-status`) or custom hooks to post data (`https://argocd-notifications.readthedocs.io/en/latest/services/webhook/#set-github-commit-status`).

We want to come up with something new, which is not easy because a lot of useful scenarios have already been covered. So, let's look at an example where we start a pipeline in GitLab CI/CD. GitLab is being used more and more these days for CI/CD because it enables pipelines where everything runs on containers in a cloud-native way with its Kubernetes runner: `https://docs.gitlab.com/runner/executors/kubernetes.html`. For our demo, we will not be using the Kubernetes executor; we will be using the shared one provided by GitLab, which is based on `docker-machine`: `https://docs.gitlab.com/runner/executors/docker_machine.html`. The pipeline we are creating will run the same on both Kubernetes and Docker machine runners.

First, create a user on GitLab by going to `https://gitlab.com/users/sign_up`. Once you have the account up and running, go ahead and create a project. There should be a **New project** button somewhere in the top-right corner of the GitLab UI. Choose **Create blank project** on the next page, after which you should be able to set the project's name.

In my case, I named it `resume-manual-pipeline` and I set the project to **Public** so that I can share it with everyone. You can set it however you wish:

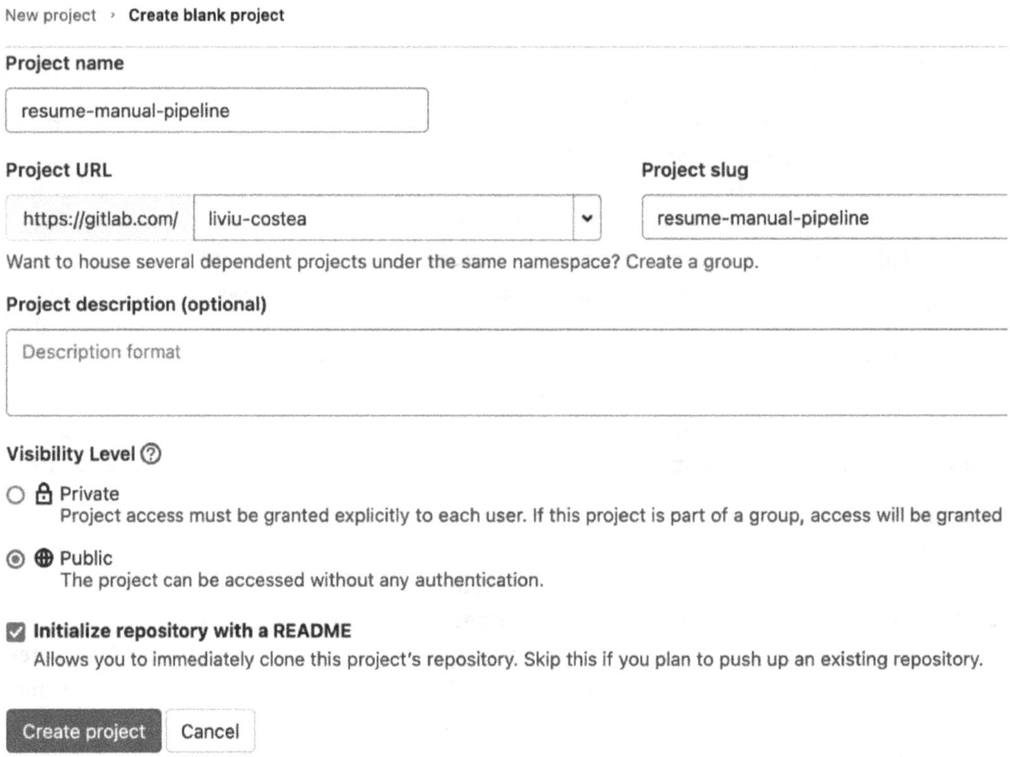

New project › **Create blank project**

Project name

resume-manual-pipeline

Project URL

https://gitlab.com/   liviu-costea

Project slug

resume-manual-pipeline

Want to house several dependent projects under the same namespace? Create a group.

Project description (optional)

Description format

Visibility Level ⓘ

○ 🔒 Private
   Project access must be granted explicitly to each user. If this project is part of a group, access will be granted

◉ 🌐 Public
   The project can be accessed without any authentication.

☑ **Initialize repository with a README**
   Allows you to immediately clone this project's repository. Skip this if you plan to push up an existing repository.

Create project   Cancel

Figure 3.3 – Creating a new GitLab project

Once we've created the project, before adding any code to it, we need to set up an easy authentication method for our Git repos with an SSH key. First, go to the top-right corner, locate the last link on the right, and click on it – you should see the **Preferences** menu item. This will take you to a page where you have a big menu on the left, including an **SSH Keys** entry. Clicking it will take you to a page where you can add your SSH key (follow *Steps 1*, *2*, and *3* in the following screenshot to get to the **SSH Keys** page). There will be a link that explains how to generate a new one. After you have created it, you can paste the **Public** one in the textbox (not the **Private** one), give it a title, and click **Add key**:

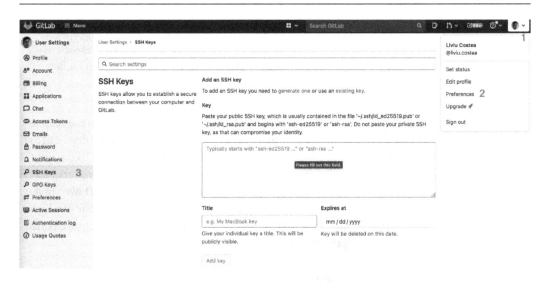

Figure 3.4 – How to get to the SSH Keys page

Now that we have the correct setup, we can clone, pull, and push our Git repos without any problems.

Now, getting back to our repo, we should clone it locally and open it in an editor. We will build a pipeline with one job named `update-deploy-status`. It will use an `alpine` Docker image and it will run a dummy script that will show the application's name, status, and the commit SHA that was applied by Argo CD. The idea is that this job can make a couple of changes, such as setting a tag to the Git commit or putting a `Production` label on some tasks after a synchronization event occurs. Ours is a dummy one to explain the link between the event and the pipeline, but yours can be more advanced. So, create a new file near the `README.md` file named `gitlab-ci.yml` and set the following pipeline definition:

```
update-deploy-status:
  stage: .post
  script:
    - echo "Deploy status $APPLICATION_DEPLOY_STATUS for
Application $APPLICATION_NAME on commit $APPLICATION_GIT_
COMMIT"
  image: alpine:3.14
  only:
    variables:
      - $APPLICATION_DEPLOY_STATUS
      - $APPLICATION_NAME
      - $APPLICATION_GIT_COMMIT
```

> **Note – Manual GitLab pipeline**
> A manual start can be done by using the condition that certain variables need to be present for the job to run (see the `only:` part). This allows us to also start the pipeline from the GitLab UI, which is a good way to debug it.

Next, we will create a commit with the `.gitlab-ci.yml` file we created and push it to the remote repo. Before we define the webhook, we need a way to authenticate the Argo CD Notifications call to our GitLab pipeline. We will use a pipeline trigger token for this: `https://docs.gitlab.com/ee/api/pipeline_triggers.html`. We will create it from the GitLab UI. On the project's main page, in the left menu, there is a **Settings** entry. Upon clicking it, you will see the **CI/CD** item in its submenu. Clicking on it will take you to a page with many sections that you can expand, one of which is **Pipeline triggers**. There, you can create a new trigger; I named mine `Argo CD Notifications Webhook`. After clicking on **Add trigger**, the token will appear:

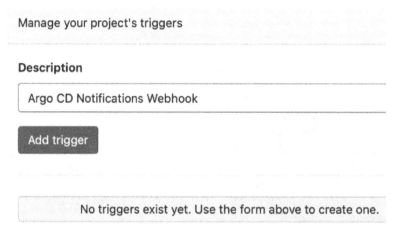

Figure 3.5 – Creating a pipeline trigger – give it a name and click the Add trigger button

We now have a token that we can use to authenticate when we want to start the pipeline from the Argo CD Notifications webhook. In the **Pipeline triggers** section, we already have an example of what the webhook should look like – all we need to do is adjust it with our configuration. TOKEN is the one we just created. In our case, REF_NAME is the `main` branch. Regarding the variables, we will populate them in the Argo CD Notifications template:

```
curl -X POST \ -F token=TOKEN \ -F "ref=REF_NAME" \ -F
"variables[RUN_NIGHTLY_BUILD]=true" \ https://gitlab.com/api/
v4/projects/29851922/trigger/pipeline
```

For the next few steps, we can follow the official documentation on how to create the webhook in Argo CD Notifications (`https://argocd-notifications.readthedocs.io/en/stable/services/webhook/`).

We need to modify the `argocd-notifications-cm` ConfigMap, which we can do by making a change in Git. In the `notifications` folder, which we created when we installed Argo CD Notifications, we need to add a new folder called `patches`. Inside this, we will add a file called `argocd-notifications-cm.yaml`, where we will define the trigger, when to send the webhook, and what the webhook should look like, which involves a notification template. We named the trigger `on-sync`. We are activating it when the sync ends up as `Succeeded`, `Error`, or `Failed` and we are linking it to the `gitlab-webhook` template. Next, the template is linked to the `gitlab` webhook, which shows that an HTTP POST request will be sent with the required variables to start our job, with `ref` set to `main`, as well as the authentication token (you will need to set this to a real value – the one that you created earlier):

```
apiVersion: v1
kind: ConfigMap
metadata:
 name: argocd-notifications-cm
data:
 trigger.on-sync: |
   - when: app.status.operationState.phase in ['Succeeded',
'Error', 'Failed']
     send: [gitlab-webhook]
 service.webhook.gitlab: |
   url: https://gitlab.com/api/v4/projects/29851922/trigger/
pipeline
   headers:
   - name: Content-Type
     value: multipart/form-data
 template.gitlab-webhook: |
   webhook:
     gitlab:
       method: POST
       body: ref=main&token=<token-goes-
here>&variables[APPLICATION_DEPLOY_STATUS]={{.app.status.
sync.status}}&variables[APPLICATION_NAME]={{.app.metadata.
name}}&variables[APPLICATION_GIT_COMMIT]={{.app.status.
operationState.operation.sync.revision}}
```

We need an entry in the `kustomization.yaml` file to reference this new file:

```
apiVersion: kustomize.config.k8s.io/v1beta1
kind: Kustomization
```

```
namespace: argocd
bases:
  - github.com/argoproj-labs/argocd-notifications/manifests/
controller?ref=v1.1.1
patchesStrategicMerge:
  - patches/argocd-notifications-cm.yaml
```

Now, create the commit, push it to remote, and make sure that the Argo CD application was synchronized to include our changes. We should now be ready to update one of our application custom resources to subscribe to the webhook we created. In this case, we have the applications in Git, but they are not tracked directly by Argo CD, so if we change them, we still need to apply them manually. In *Chapter 5, Argo CD Bootstrap K8s Cluster*, we will look at the app-of-apps pattern, which allows us to store all the application definitions in Git. For now, though, we are OK with performing these small changes manually.

Earlier, in the ch03 folder, we created an argocd-app.yaml file. Here, we are going to modify it so that it includes the annotation that will specify that it is subscribed to the gitlab webhook with the on-sync trigger (see the highlighted code):

```
apiVersion: argoproj.io/v1alpha1
kind: Application
metadata:
  annotations:
    notifications.argoproj.io/subscribe.on-sync.gitlab: ""
  name: argocd
spec:
  destination:
    namespace: argocd
    server: https://kubernetes.default.svc
  project: default
  source:
    path: ch03/kustomize-installation
    repoURL: https://github.com/PacktPublishing/ArgoCD-in-
Practice.git
    targetRevision: main
```

We need to apply this file manually with the following command:

```
kubectl apply -f argocd-app.yaml
```

The output should look like this:

```
application.argoproj.io/argocd configured
```

With that, we've done all the setup so that every change and synchronization we make on the `argocd` application will start a pipeline in our GitLab project. Here, we can modify the `argocd-cm` ConfigMap we added earlier in this chapter when we discussed Argo CD self-management. After we push the change to remote, we should have a pipeline that provides an output similar to the following:

```
32  Using docker image sha256:14119a10abf4669e8cdbdff324a9f9605d99697215a0d21c360fe8dfa8471bab for alpine:3.14 with digest
    2e3d3c45cccac829840a25941e679c25d438cc8412c2fa221cf1a824e6a ...
33  $ echo "Deploy status $APPLICATION_DEPLOY_STATUS for Application $APPLICATION_NAME on commit $APPLICATION_GIT_COMMIT"
34  Deploy status Synced for Application argocd on commit e8035777f082ff4d214503d0a0b26a2281b5a0f7
36  Cleaning up project directory and file based variables
38  Job succeeded
```

Figure 3.6 – The GitLab job's output from the pipeline that was started by Argo CD Notifications

In this section, we went through quite a lengthy demo where we created a small GitLab pipeline with one job that gets triggered with a notification when a failed or successfully performed synchronization occurs in an Argo CD application. The other option would have been to regularly query the application sync status from a pipeline when a new commit was performed until it gets to the state we are waiting for and then have to perform the actions we need. These notifications allow us to embrace the full pull nature of GitOps and not have to create workarounds that make it look like a push method.

Summary

We started this chapter by installing Argo CD with Kustomize. We went with a cloud provider-managed cluster because we needed more nodes to experiment with an HA deployment. We experienced how Argo CD can update itself and how we can make configuration changes to our installation. While a production Kubernetes cluster is highly available and a cloud provider will manage it for us, there are still scenarios when disasters can happen, so we need to have a working disaster recovery strategy. We saw how we can create backups of the Argo CD state and then restore it in a new cluster.

Observability is an important topic and we discussed which metrics can be used to monitor an Argo CD installation, from the OOM container restarts to what a microservice team needs to watch out for. We finished by learning how to link the result of a synchronization to a pipeline so that everything will be automated.

In the next chapter, we are going to discover how we can bootstrap a new Kubernetes cluster in AWS using Argo CD, including how to set up applications in the newly created cluster, such as external DNS and Istio.

Further reading

To learn more about the topics that were covered in this chapter, take a look at the following resources:

- The Argo project's official blog (this includes Argo CD and other projects from the Argo organization): `https://blog.argoproj.io`.

- For an HA installation, it is also important that the Pods are spread across nodes (and availability zones). Go to `https://kubernetes.io/docs/concepts/scheduling-eviction/assign-pod-node/` to see how this is done with node affinity.

- Observability with Prometheus: `https://prometheus.io/docs/introduction/overview/`.

4

Access Control

In this chapter, we're going to discover how to set up a user's access to Argo CD, what options we have in order to connect with the CLI from a terminal or a CI/CD pipeline, and how **role-based access control** is performed. We will be checking out the **single sign-on** (**SSO**) options, which usually is a feature that you need to pay for, but with Argo CD being completely open source and without a commercial offering, you receive that out of the box.

The main topics we are going to cover are as follows:

- Declarative users

- Service accounts

- Single sign-on

Technical requirements

For this chapter, we assume you already have an installation of Argo CD up and running on your device or on a cloud platform. We don't need a highly available one; it can be in a local Kubernetes cluster, but we do need one that can manage itself because we are going to make several changes to it. For the single sign-on part, we will need an installation that can be reached by our external provider, so we will need a domain name or a public IP for it.

We are going to write some code (YAML to be more precise), so a code editor will be needed as well. I am using Visual Studio Code (https://code.visualstudio.com). All the changes we will be making can be found at https://github.com/PacktPublishing/ArgoCD-in-Practice in the ch04 folder.

Declarative users

The **Open Web Application Security Project** (**OWASP**) – https://owasp.org – is a nonprofit foundation that does a lot of work with regard to web application security. Their most well-known project is the OWASP Top Ten (https://owasp.org/www-project-top-ten/), which is a list of the most important risks to consider when it comes to the security of our web applications. They update this list every few years. Currently, in their latest version, which is from 2021, so brand new, we have in first place **Broken Access Control** (it was in the fifth position on the previous 2017 list). Since the primary goal of the Top Ten is to bring awareness in the community about the major web security risks, we can understand with Broken Access Control being at the top that it is critical to do a proper setup for our users and the kind of access everyone gets in order to not violate the principle of least privilege. It would be common for our development teams to get write access to development environments and read-only for our production ones, while an SRE would have full rights. Next, we will see how to accomplish that.

Admin and local users

After we install Argo CD in the cluster, all we get is the admin user. Initially, the password for the admin user was the name of the application server Pods (and if there were more, then the first one that started); it looked something like this – *argocd-server-bfb77d489-wnzjk*. But this was considered a poor choice because the first part (the Pod starts with *argocd-server-*) was fixed for most installations, while the second part was generated and it was supposed to have random characters, but they weren't actually random (more details can be found at `https://argo-cd.readthedocs.io/en/stable/security_considerations/#cve-2020-8828-insecure-default-administrative-password`). Moreover, it was visible to everyone who had cluster access (which didn't necessarily mean Argo CD admin access). So, ever since version 2.0.0 was released, a new password is generated for the user when the application is installed and it is saved in a Secret called `argocd-initial-admin-secret`. We can see what the password is after installation with this command:

```
kubectl -n argocd get secret argocd-initial-admin-secret -o
jsonpath="{.data.password}" | base64 -d
```

The command output is your password and it should look something like this (from which you should remove the % character because the shell inserted this CR/LF in order for the prompt to start on a new line):

```
pOLpc19ah90dViCD%
```

Now, you can use the password for the UI and CLI to check that it all works. For updating the password, you can use the UI by navigating to the **User-Info** section, or there is also the possibility of doing it via the CLI with the `argocd account update-password` command (`https://argo-cd.readthedocs.io/en/stable/user-guide/commands/argocd_account_update-password/`). If you forget the password, there is a way to reset it and it involves making some changes directly on the Argo CD main Secret resource where the bcrypt hashes are stored – more details can be found at `https://argo-cd.readthedocs.io/en/stable/faq/#i-forgot-the-admin-password-how-do-i-reset-it`.

The next thing after we have the admin user up and running would be to disable it because it is just too powerful and we want to follow the least privilege principle, which means we should use only the minimal type of access that we need in order to accomplish our work (learn more about it here: `https://en.wikipedia.org/wiki/Principle_of_least_privilege`). But in order to not lose access to the system, we need to first create a local user with fewer rights that will allow us to perform our day-to-day tasks. We will name it `alina` and allow it access to both the UI and the CLI. The rule of thumb here is to always have a dedicated user for every one of your colleagues who needs to access Argo CD instead of using it as a group or a team user. This helps if somebody loses access to the system as they are leaving the team because we can always disable or remove their account. To add a new user, the ConfigMap `argocd-cm` file should look like this:

```
apiVersion: v1
kind: ConfigMap
metadata:
  name: argocd-cm
data:
  accounts.alina: apiKey, login
```

We should create the commit and push this update to remote so that our configuration change is applied by Argo CD (in order for this ConfigMap to be applied automatically when the push is done, you need to have Argo CD installed and configured as described in *Chapter 3, Operating Argo CD*). Assuming that the CLI is installed and points to our API server instance (the login being done with the admin user), we can verify that the new user is created by running the following code:

```
argocd account list
```

The output should be like this:

```
NAME     ENABLED   CAPABILITIES
admin    true         login
alina    true         apiKey, login
```

This means that we have the user ready, but we still need to set its password. For this, we can use this command, using the admin password as the current password:

```
argocd account update-password --account alina --current-
password pOLpcl9ah90dViCD --new-password k8pL-xzE3WMexWm3cT8tmn
```

For that, the output is as follows:

```
Password updated
```

When running this `update-password` command, it is possible to skip the passing of the `--current-password` and `--new-password` parameters; this way, their values will not be saved in the shell history. If they are not passed, you will be asked to introduce them in an interactive way, which is a much safer choice.

To see what happened, we need to take a look at the `argocd-secret` Secret resource by running this command. We will see new entries there for our user, just like for the admin user:

```
kubectl get secret argocd-secret -n argocd -o yaml
```

And we should get the following (we will remove some of the fields that are of no interest to us):

```
apiVersion: v1
data:
  accounts.alina.password: JDJhJDEwJHM0MVdJdTE5UEFZak9oUD
dWdk9iYS5KVEladjFjLkdReFZ0eVdlN0hYLnlLem01Z3BEelBX
  accounts.alina.passwordMtime: MjAyMS0xMC0wOVQxMzo1MDoxNVo=
  accounts.alina.tokens: bnVsbA==
  admin.password: JDJhJDEwJEd1LzlZVG4uOEVmY2x3bGV6dDV3amVDd2d
jQW1FUEk3c21BbGoxQjV0WXXIxek9pYUxjL1ZL
  admin.passwordMtime: MjAyMS0xMC0wOFQwNzo0Mzo1NVo=
  server.secretkey: WkI3dzBzYkpoT1pnb2Njb2ZZSEh0NmppQUFM3YkRDMTN
zc1Z5VWpCNWRKcz0=
kind: Secret
metadata:
  labels:
    app.kubernetes.io/instance: argocd
    app.kubernetes.io/name: argocd-secret
    app.kubernetes.io/part-of: argocd
  name: argocd-secret
  namespace: argocd
type: Opaque
```

If we look at the tokens created for this user (in the `accounts.alina.tokens` field), we will see that it is actually null, which means that we don't have any created for now:

```
echo "bnVsbA==" | base64 -d
```

That gives the following output:

```
null%
```

If you are wondering about what rights this user has, it actually can't do anything yet. We can log in with it, but if we try to list applications or clusters, we will get empty lists. In order to allow it to do something, we have two options: either we will give specific rights to the user, or we set the default policy that every user will fall back to when authenticated in case nothing specific for it can be found. We are going to set the default policy to read-only now and we will check how to add specific rights when we use the access token. In order to do this, there is a new ConfigMap resource we need to modify, which is where we will set all our **Role-Based Access Control** (**RBAC**) rules. Create a new file named `argocd-rbac-cm.yaml` in the same location where we have the `argocd-cm.yaml` ConfigMap and add this content:

```
apiVersion: v1
kind: ConfigMap
metadata:
  name: argocd-rbac-cm
data:
  policy.default: role:readonly
```

Don't forget to add an entry to the `kustomization.yaml` file. Now, under `patchesStrategicMerge`, we should have two entries for both the ConfigMaps we modified:

```
patchesStrategicMerge:
  - patches/argocd-cm.yaml
  - patches/argocd-rbac-cm.yaml
```

Create the commit for both the files and push it to remote so that Argo CD can apply the RBAC default policy. After the change is made, we can use the new user with the CLI. To do this, first we should log in to our API server (in my case, I used port-forwarding from my local installation so the server is on `https://localhost:8083`):

```
argocd login localhost:8083 --username alina
```

You will probably receive a warning message that the certificate comes from an unknown authority, which is correct because it is self-signed. We can ignore it for now because we are connecting to localhost:

```
WARNING: server certificate had error: x509: certificate signed
by unknown authority. Proceed insecurely (y/n)?
```

Next, enter the password for the `alina` account and you should be logged in and ready to issue commands. We can try to list the Applications Argo CD has installed to make sure we have the read access:

```
argocd app list
```

The output should be as follows (before setting the default policy to read-only, this list would have been empty):

```
CURRENT   NAME                   SERVER
*         localhost:8083         localhost:8083
```

Now, we can move on and disable the admin user, for which we can use the `admin.enabled` field and set it to `false` in the `argocd-cm` ConfigMap. It should look like this:

```
apiVersion: v1
kind: ConfigMap
metadata:
 name: argocd-cm
data:
 accounts.alina: apiKey, login
 admin.enabled: "false"
```

Commit the changes and push them to remote, so that the admin user can't be used anymore. You just moved one step further toward securing your Argo CD instance and getting it ready for production.

We have seen how we can deal with the users in a declarative way, how to create new local ones and disable the admin, and how passwords are handled. Next, we will learn about the users used in automations, the so-called service accounts.

Service accounts

Service accounts are the accounts we use for authenticating automations such as CI/CD pipelines to the system. They should not be tied to a user because we don't want our pipelines to start failing if we disable that user, or if we restrict its rights. Service accounts should have strict access control and should not be allowed to do more than what is required by the pipeline, while a real user will probably need to have access to a larger variety of resources.

There are two ways to create service accounts in Argo CD: one is with local users (for which we only use `apiKey` and remove the `login` part) and the other is to use project roles and have tokens assigned for those roles.

Local service accounts

We are now going to create a separate local account that only has the `apiKey` functionality specified. This way, the user doesn't have a password for the UI or the CLI and access can be accomplished only after we generate an API key for it (which gives it CLI or direct API access). We will also set specific rights for it; for example, we will allow it to start a sync for specific applications.

We will modify the `argocd-cm` ConfigMap to add the new service account user, which we will name `gitops-ci`:

```
apiVersion: v1
kind: ConfigMap
metadata:
  name: argocd-cm
data:
  accounts.alina: apiKey, login
  accounts.gitops-ci: apiKey
  admin.enabled: "false"
```

Commit and push it to remote so that we have the new account up and running. After we have the new account created, we need to run a command in order to generate an access token. The issue here is that the `alina` user doesn't have the rights to do this, and the admin account was disabled. We can re-enable admin, but that's not a good practice because we could always forget to disable it again, or just leave it enabled for a long time. Normally, we should disable admin only after we finish the setup of all the local users. However, we might need to create new ones at any point because there are either new people joining our teams, or new scenarios to be automated with pipelines.

So, let's see how we can assign the rights to update accounts to the user `alina`. To do this, we will modify the `argocd-rbac-cm.yaml` file to create a new role for user updates named `role:user-update` and we will assign the role to the user (the lines that are used to define the policies start with p, while the lines to link users or groups to roles start with g):

```
apiVersion: v1
kind: ConfigMap
metadata:
 name: argocd-rbac-cm
data:
  policy.default: role:readonly
  policy.csv: |
    p, role:user-update, accounts, update, *, allow
    p, role:user-update, accounts, get, *, allow
    g, alina, role:user-update
```

> **Note – RBAC – policy.csv**
>
> If you are wondering where the `policy.csv` file is coming from, or what is with p and g and all these notations, you can check out the `casbin` library that Argo CD is using to implement RBAC: `https://casbin.org/docs/en/get-started`.

Commit, push the changes to remote, and make sure they are applied by Argo CD. Now, we can run the command in order to create the token (we assume that you are logged in with the CLI with the user `alina`):

```
argocd account generate-token -a gitops-ci
```

The output should be similar to this (and it is the account token):

```
eyJhbGciOiJIUzI1NiIsInR5cCI6IkpXVCJ9.eyJqdGkiOiIzZTc2NWI5Ny04MG
YyLTRkODUtYTkzYi1mNGIzMjRkYTc0ODciLCJpYXQiOjE2MzQxNDkyODksImlzc
yI6ImFyZ29jZCIsIm5iZiI6MTYzNDE0OTI4OSwic3ViIjoiZ2l0b3BzLWNpOmFw
aUtleSJ9.azbvrvckSDevFOG6Tun9nJV0fEMcMpI9Eca9Q5F2QR4
```

One easy way to check whether the newly generated token is working is by running the `argocd account get-user-info` command. We can run it once with the logged-in user and again with the authentication token. Here, I am putting the example only for the token (for the logged-in user, just don't pass the `--auth-token` parameter). The `--auth-token` flag is a general one, meaning it can be used for any command (and of course you will have another token, so don't forget to replace it):

```
argocd account get-user-info --auth-
token eyJhbGciOiJIUzI1NiIsInR5cCI6IkpXVCJ9.eyJqdGkiOiIzZTc2NWI5
Ny04MGYyLTRkODUtYTkzYi1mNGIzMjRkYTc0ODciLCJpYXQiOjE2MzQxNDkyODk
sImlzcyI6ImFyZ29jZCIsIm5iZiI6MTYzNDE0OTI4OSwic3ViIjoiZ2l0b3BzLW
NpOmFwaUtleSJ9.azbvrvckSDevFOG6Tun9nJV0fEMcMpI9Eca9Q5F2QR4
```

The output should be similar to this:

```
Logged In: true
Username: gitops-ci
Issuer: argocd
Groups:
```

For the newly created service account, since we didn't specify any permissions, it'll take the default ones, meaning the read-only access. Of course, creating such a token with just read access doesn't make a lot of sense. Normally, it should do something like register a new cluster or unregister it (which is usually done in an imperative way – by running a command), create or remove users, create an application, synchronize it, and so on. We are not able to set a local account to be a part of an RBAC group, we can only have roles and assign local users to the roles. We will see how groups work when we talk about SSO users later in this chapter. For a complete list of resources and actions you can use for your local accounts in the RBAC ConfigMap, you can check the official documentation: `https://argo-cd.readthedocs.io/en/stable/operator-manual/rbac/#rbac-resources-and-actions`.

Project roles and tokens

Project roles are the second option we can use for service accounts. An application project is a way for us to apply some constraints on the application definition. We can set the repository from where we take the state, the destination cluster, and the namespaces where we can deploy and even filter the resource types that we can install (for example, we can declare that the applications that use this project can't deploy Secrets). Besides these, there is also the possibility to create a project role and to specify the kind of actions it is allowed to do, and a role can then have tokens assigned.

When Argo CD is installed, it also comes with a default project, actually called default, that doesn't set any restrictions to its applications (everything is set to allow: '*'). In order to show how projects are used with their tokens, we will create a new project and use it for our existing *argocd* application. Once we have it, we will need to create a role for the project, assign permissions to the role, and create a token.

We will create a new file called argocd-proj.yaml and we will store it in the same location as our argo-app.yaml file (the files can also be found in our official repo at https://github.com/PacktPublishing/ArgoCD-in-Practice in the ch04/kustomize-installation folder). After we create it, we will need to apply it manually (we will see in *Chapter 5, Argo CD Bootstrap K8s Cluster*, how the app-of-apps pattern can help us create all these Applications and AppProjects automatically). This is the content of the file for now. You can see the restrictions for the destination namespace and cluster and also that we can only load the state from a fixed repo. We also added a new role called *read and sync*, for which we allowed the *get and sync* actions for all the applications under the AppProject:

```
apiVersion: argoproj.io/v1alpha1
kind: AppProject
metadata:
  name: argocd
spec:
  roles:
    - name: read-sync
      description: read and sync privileges
      policies:
        - p, proj:argocd:read-sync, applications, get,
argocd/*, allow
        - p, proj:argocd:read-sync, applications, sync,
argocd/*, allow
  clusterResourceWhitelist:
    - group: '*'
      kind: '*'
```

```
    description: Project to configure argocd self-manage
application
    destinations:
    - namespace: argocd
      server: https://kubernetes.default.svc
    sourceRepos:
    - https://github.com/PacktPublishing/ArgoCD-in-Practice.git
```

We will apply the AppProject to the cluster with `kubectl apply`:

```
kubectl apply -n argocd -f argocd-proj.yaml
```

The command output should be similar to this:

```
appproject.argoproj.io/argocd created
```

Then, we will modify the `argocd` Application to start using this new AppProject, changing from `project: default` to `project: argocd` in the definition. Now, apply the updated file to the cluster with `kubectl apply` (the complete `argocd-app.yaml` file can be found at `https://github.com/PacktPublishing/ArgoCD-in-Practice` in `ch04/kustomize-installation`):

```
kubectl apply -n argocd -f argocd-app.yaml
```

Now, we should be ready to generate our token for the read and sync role we created. This needs to be done in a synchronous way, either with the CLI or from the UI, so we can retrieve the generated token. We can run the CLI command using the user `alina`, but we will get an error as it doesn't have the needed permissions:

```
FATA[0000] rpc error: code = PermissionDenied desc =
permission denied: projects, update, argocd, sub: alina, iat:
2021-10-16T10:16:33Z
```

Let's add the needed permissions by modifying the RBAC ConfigMap. I am pasting just the `policy.csv` part here and, this time, I will restrict the action only to our exact project, named `argocd`:

```
policy.csv: |
    p, role:user-update, accounts, update, *, allow
    p, role:user-update, accounts, get, *, allow
    p, role:user-update, projects, update, argocd, allow

    g, alina, role:user-update
```

Don't forget to commit and push it to remote so that our change will be picked up by Argo CD and applied to the cluster. After this, we can generate the token using the following command:

```
argocd proj role create-token argocd read-sync
```

The response will look something like this:

```
Create token succeeded for proj:argocd:read-sync.
  ID: ccdc5906-11fc-483b-8e8d-0511c6f28978
  Issued At: 2021-10-16T13:44:19+03:00
  Expires At: Never
  Token: eyJhbGciOiJIUzI1NiIsInR5cCI6IkpXVCJ9.eyJqdGkiOiJjY2RjN
TkwNi0xMWZjLTQ4M2ItOGU4ZC0wNTExYzZmMjg5NzgiLCJpYXQiOjE2MzQzODEw
NTksImlzcyI6ImFyZ29jZCIsIm5iZiI6MTYzNDM4MTA1OSwic3ViIjoicHJvajp
hcmdvY2Q6cmVhZC1zeW5jIn0.R02VHylpb4aPjtpd5qLOXHpELGOVgnelCJr3q8
bGU5Y
```

I can now use this token to start a manual sync for all the Applications under the `argocd` AppProject. We only have one Application and it is named the same as the AppProject, `argocd`. We can see that if we try to call the sync using the logged-in user `alina`, we will get a permissions-denied error, as can be seen in the output following this command:

```
argocd app sync argocd
```

The error will be as follows:

```
FATA[0000] rpc error: code = PermissionDenied desc = permission
denied: applications, sync, argocd/argocd, sub: alina, iat:
2021-10-16T10:16:33Z
```

Using the generated token, the command becomes this:

```
argocd app sync argocd --auth-token eyJhbGciOiJIUzI1NiIsInR5cCI
6IkpXVCJ9.eyJqdGkiOiJjY2RjNTkwNi0xMWZjLTQ4M2ItOGU4ZC0wNTExYzZmM
jg5NzgiLCJpYXQiOjE2MzQzODEwNTksImlzcyI6ImFyZ29jZCIsIm5iZiI6MTYz
NDM4MTA1OSwic3ViIjoicHJvajpoicHJvajphcmdvY2Q6cmVhZC1zeW5jIn0.R02VHylpb4a
Pjtpd5qLOXHpELGOVgnelCJr3q8bGU5Y
```

The output is very long because it starts to enumerate all the resources our Argo CD Application has installed along with their status and this means that the sync is working. It can be seen also from the sync status in the UI (go to the **argocd Application** and on its page, you should have a **Sync Status** button, which shows details about the last initiated sync):

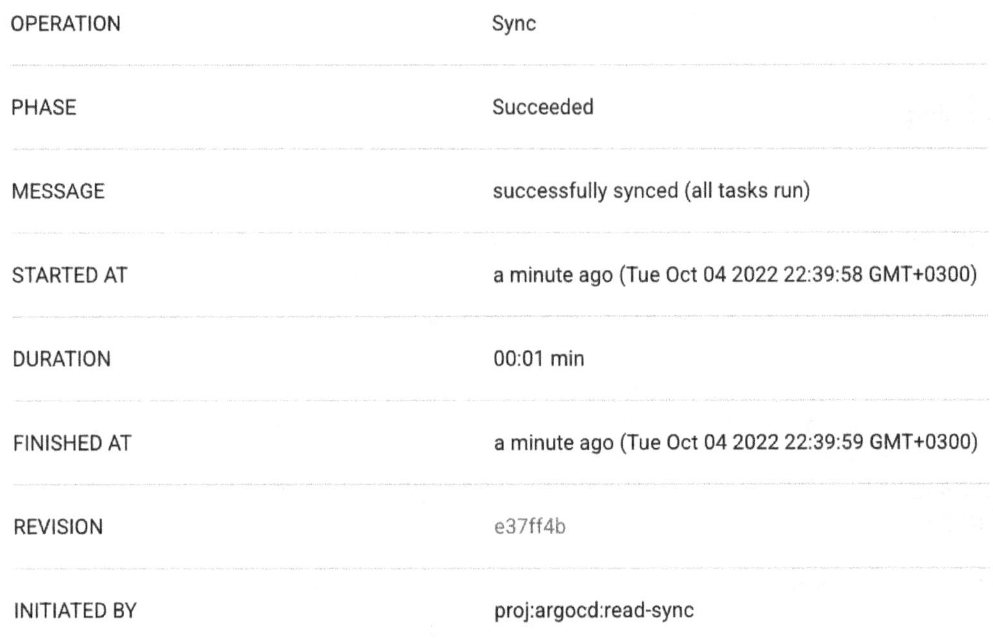

OPERATION	Sync
PHASE	Succeeded
MESSAGE	successfully synced (all tasks run)
STARTED AT	a minute ago (Tue Oct 04 2022 22:39:58 GMT+0300)
DURATION	00:01 min
FINISHED AT	a minute ago (Tue Oct 04 2022 22:39:59 GMT+0300)
REVISION	e37ff4b
INITIATED BY	proj:argocd:read-sync

Figure 4.1 – argocd: Sync initiated by the project role

Every token that we generate is saved into the project role. We can check the amount of time we used it for and whether it is about time to rotate it. We can also set an expiration date for it if we plan to use it for a limited time. Otherwise, putting a hard deadline for when we will need to refresh it will require strong discipline from the engineers managing Argo CD. In my case, I usually have a tendency to postpone the quarterly review of access control by up to a week or two, so having expiration dates might lead to failed pipelines.

> **Note – Tokens with sync action only**
>
> Does it make sense to have tokens (either coming from local accounts or project roles) that only do the sync action as long as we can allow Applications to sync automatically? I think so, and a valid scenario would be to still allow Applications to sync automatically but set a bigger timeout, such as 10-15 minutes (which can reduce the system load and increase its performance). However, after we make a commit to the GitOps repo, we can call the sync from the pipeline as well.

Project roles can only be used for performing actions on Applications' resources and those Applications need to be under the project for which the role is created, so it is pretty limited in scope. However, it is actually such constraints that transform them into good candidates to use in our pipelines because if they get compromised for any reason, then the attack surface is a reduced one.

Single sign-on

Single Sign-On (**SSO** – `https://en.wikipedia.org/wiki/Single_sign-on`) allows us to have a master login, and based on that, we can be authorized into other independent applications (they do keep a connection with the master system even if they are independent). It means that if I want to access `argocd.mycompany.com`, then `argocd.mycompany.com` will trust an external provider in order to verify my identity. Furthermore, the type of access I will receive to `argocd.mycompany.com` can be controlled from the external master system based on the groups I am part of, or sometimes by setting attributes on the account.

I know that there are companies that see SSO as mandatory; they will not adopt a new tool if it doesn't offer the option. And it does make sense considering it has security benefits, such as not needing passwords for all the applications we use, and ease of onboarding/offboarding colleagues by controlling everything from one dashboard. These factors rank this feature as a must-have one.

Luckily, Argo CD offers the SSO feature with two ways of configuring it. One is by using the Dex OIDC provider (`https://github.com/dexidp/dex`) that is installed by default, and the second is directly with Argo CD (skipping the Dex installation) when you rely on another OIDC provider. We will look at examples for both these cases.

Argo CD offers SSO logins for the UI and for the CLI. Once you have SSO enabled and if there are no local accounts that can log in and admin is disabled, the user/password form will be automatically removed from the UI and only the log in via SSO button will remain.

Figure 4.2 – No user/password login with SSO on and admin disabled

For the CLI, the log in with SSO command will look like this (replace `argocd.mycompany.com` with your correct server address):

```
argocd login --sso argocd.mycompany.com
```

By using SSO, you significantly reduce the attack surface because only one set of credentials is needed to get access to many applications. And in Argo CD, SSO is well covered and it can be accomplished with the help of Dex, or with Argo CD directly when using an external OIDC provider. Next, we are going to explore both options, starting with Dex.

SSO with Dex

Dex is an identity system that allows applications to offload authentication. Dex is able to use different authentication schemes such as LDAP or SAML and can connect to well-known identity providers, such as Google, GitHub, and Microsoft. The main advantage is that you connect once with Dex and then get the benefit of all the connectors it supports (for example, you could have authentication to your website with Google, Microsoft, and GitHub – all just by using Dex).

The official documentation of Argo CD covers this topic very well – `https://argo-cd.readthedocs.io/en/stable/operator-manual/user-management/#sso`. There are extensive examples on how to configure Google via SAML (`https://argo-cd.readthedocs.io/en/stable/operator-manual/user-management/google/#g-suite-saml-app-auth-using-dex`), Microsoft Azure Active Directory (`https://argo-cd.readthedocs.io/en/stable/operator-manual/user-management/microsoft/#azure-ad-saml-enterprise-app-auth-using-dex`), or GitHub (`https://argo-cd.readthedocs.io/en/stable/operator-manual/user-management/#2-configure-argo-cd-for-sso`).

> **Dex SAML Connector – Unmaintained**
>
> At the time of finalizing this book (October 2022), the Dex SAML connector is proposed to be deprecated because it is not maintained anymore. Still, this is the main way to connect to providers, so we hope it will get into better shape in the future and that's why our demo is based on it. For more details, please follow `https://github.com/dexidp/dex/discussions/1884`.

RBAC groups

When we use SSO, a user can be added automatically to our RBAC groups without any configuration per user on the Argo CD side. Everything will be controlled by the SSO system.

When we use Google SAML, which is pretty popular, we don't have the option to export the group information that we set in the Google system. However, we do have a way to expose additional user details from Google and that is by using a custom attribute, based on which we can assign the user to an Argo CD RBAC group. The custom attribute needs to be defined initially and then set its value for each user. In order to do all these changes, including the ones from the initial setup, you need administrative access for your GSuite account.

Do follow the official tutorial for the Google SAML setup initially before we continue with the group's configuration. Now, let's define the additional schema for the attribute, which we will name ArgoCDSSO, while we name the role ArgoCDRole. Go to `https://developers.google.com/admin-sdk/directory/reference/rest/v1/schemas/insert?authuser=1` and use the *Try this API* form. Set the `my_customer` value for the `customerId` field (which means to use the customer ID of the logged-in user) and in `Request Body`, add the following JSON, which defines our new attribute:

```
{
    "schemaName": "ArgoCDSSO",
    "displayName": "ArgoCD_SSO_Role",
    "fields": [
        {
            "fieldType": "STRING",
            "fieldName": "ArgoCDRole",
            "displayName": "ArgoCD_Role",
            "multiValued": true,
            "readAccessType": "ADMINS_AND_SELF"
        }
    ]
}
```

After we execute the form, we can go back to the *SAML Attribute Mapping* page so that we can add the additional attribute. We will use the `name` role for the mapping, while in the first dropdown, we will be able to select the newly created `ArgoCDSSO` schema, and in the second dropdown, we will choose the `ArgoCDRole` attribute.

In the `argocd-cm` ConfigMap where we added the `dex.config` section, we will need to make a small change in order to introduce the mapping between the RBAC groups and the role we created in Google SAML. I am pasting only the part with `dex.config` from the file. What is important is the last line – the value of `groupsAttr` needs to match the name of the mapping (the complete `argocd-cm.yaml` file can also be found in the official repository for the book: `https://github.com/PacktPublishing/ArgoCD-in-Practice` in ch04/sso-setup):

```
url: https://argocd.mycompany.com
dex.config: |
  connectors:
```

```
    - type: saml
        id: argocd-mycompany-saml-id
        name: google
        config:
        ssoURL: https://accounts.google.com/o/saml2/
idp?idpid=<id-provided-by-google>
        caData: |
          BASE64-ENCODED-CERTIFICATE-DATA
        entityIssuer: argocd-mycompany-saml-id
        redirectURI: https://argocd.mycompany.com/api/dex/
callback
        usernameAttr: name
        emailAttr: email
        groupsAttr: role
```

We saw how RBAC roles work when we created the local account and assigned it *update user* rights via an update-user role. In the case of local accounts (no matter whether we're using username/password or an API key), we need to link each of them to a role if we want to give them rights; but for SSO, we will see they are linked automatically to groups.

Based on the tasks engineers perform, we can separate them into developers and site reliability people. This means it would make sense to have at least these two groups defined in your system. There could be more if we want to give fewer permissions for new joiners than for developers – for example, no write access, not even for the development environment applications.

When Argo CD is installed, it comes with two RBAC roles already set up, the *admin* one, which has full rights, and the *readonly* one, which, as the name suggests, can do all the read actions and no changes or deletions (if you want to see the permissions of the built-in roles, check out `https://github.com/argoproj/argo-cd/blob/master/assets/builtin-policy.csv`). The *readonly* one would be a good role for new joiners, while for developers, besides the read access, we could allow them to sync applications. SREs should be able to perform additional updates to Applications and AppProjects and even delete them. So, let's see how we can define these groups in Argo CD RBAC ConfigMap and tie them to SSO.

We will use the *readonly* role and have the new *developer* one inherit all its permissions by using such an entry `- g, role:developer, role:readonly`. Then, we will add the additional permissions we want the *developer* role to have, which in our case is syncing applications. After we finish with the *developer* role, we set the *sre* one to inherit all the permissions from *developer* and continue with adding any new permissions. In the end, it is important to make the mapping between the name of the Argo CD RBAC group and the value of the special attribute we defined in the identity provider. So, in our case, for a user to have the *developer* permissions, they should have *developer* set as the value of the `ArgoCDRole` attribute in their Google dashboard. Here is what the changes will look like in the `argocd-rbac-cm.yaml` RBAC ConfigMap. Also, this time, we didn't use the default read-only policy (a reference to this file can be seen in the official repo at `https://github.com/PacktPublishing/ArgoCD-in-Practice` in the `ch04/sso-setup` folder):

```
apiVersion: v1
kind: ConfigMap
metadata:
 name: argocd-rbac-cm
data:
 policy.csv: |

    g, role:developer, role:readonly
    p, role:developer, applications, sync, */*, allow


    g, role:sre, role:developer


    p, role:sre, applications, create, */*, allow
    p, role:sre, applications, update, */*, allow
    p, role:sre, applications, override, */*, allow
    p, role:sre, applications, action/*, */*, allow
    p, role:sre, projects, create, *, allow
    p, role:sre, projects, update, *, allow
    p, role:sre, repositories, create, *, allow
    p, role:sre, repositories, update, *, allow


    g, Sre, role:sre
    g, Developer, role:developer
    g, Onboarding, role:readonly
```

This would be a good start for defining your access control policy in Argo CD, but do expect that changes will be needed over time. You might need to give more permissions to SRE people in order to solve some strange situations in production, or provide some of the senior developers with more permissions by creating a new role and group.

If you are wondering how to allow more rights for your users based on the environments they are going to modify, there are a few ways it can be done. So, for example, we could allow SREs to have admin rights in development/staging and only some of the permissions in pre-production or production. Probably, the best way would be to have separate Argo CD instances that would take care of development and live environments. Another way is that you can use specific prefixes for Applications and AppProjects, though it might get more complicated for clusters, repositories, or accounts.

SSO with Argo CD directly

Also, for the SSO done directly with Argo CD (so, by skipping Dex), we have extensive documentation. There are many tutorials in the official documentation on how to set up external OIDC providers, including these ones:

- OKTA – `https://argo-cd.readthedocs.io/en/stable/operator-manual/user-management/okta/#oidc-without-dex`

- OneLogin – `https://argo-cd.readthedocs.io/en/stable/operator-manual/user-management/onelogin/`

- OpenUnison – `https://argo-cd.readthedocs.io/en/stable/operator-manual/user-management/openunison/`

- Keycloak – `https://argo-cd.readthedocs.io/en/stable/operator-manual/user-management/keycloak/`

The explanations are good, and they have many screenshots on how to do all the preparations on the OIDC provider side and configuration examples for Argo CD, so it is not worth repeating such tutorials in this chapter. One thing to note would be that we can make these SSO configurations more secure as we can extract the sensitive information from the ConfigMap and store it in Secret resources populated with an external operator.

OpenID Connect – OpenID Foundation

OpenID Connect was created in a working group that was part of the OpenID Foundation. The OpenID Foundation is a non-profit organization that takes care of the OpenID technologies: `https://openid.net/foundation/`.

We will take the example of OneLogin. After everything is created on the OneLogin side, the Argo CD configuration file looks like this:

```
apiVersion: v1
kind: ConfigMap
metadata:
 name: argocd-cm
data:
 url: https://argocd.mycompany.com
 oidc.config: |
```

```
name: OneLogin
issuer: https://mycompany.onelogin.com/oidc/2
clientID: aaaaaaaa-aaaa-aaaa-aaaa-aaaaaaaaaaaaaaaa
clientSecret: abcdef123456
requestedScopes: ["openid", "profile", "email", "groups"]
```

An improvement for this would be to move the `clientSecret` entry into a Secret and reference it from the ConfigMap. Referencing a Secret from the ConfigMap is not something supported by Kubernetes, but instead Argo CD will understand the notation and will read the value from the Secret. You can find out more at `https://argo-cd.readthedocs.io/en/stable/operator-manual/user-management/#sensitive-data-and-sso-client-secrets`. We want to use a separate Secret because we don't want to commit the sensitive values directly to Git. There are some solutions where we can create a **custom resource** that will know to populate the sensitive information from a different source. We have tools such as **ExternalSecrets** (`https://github.com/external-secrets/external-secrets`), **aws-secret-operator** (`https://github.com/mumoshu/aws-secret-operator`), or even a plugin for Argo CD (`https://github.com/IBM/argocd-vault-plugin`) to help us accomplish this. The **ExternalSecrets** project has some advantages because of the wide range of data stores it supports, such as **AWS Parameter Store**, **Azure Key Vault**, **Google Cloud Secrets Manager**, **Hashicorp Vault**, and some others, which means it is likely that it will fit your infrastructure. It can be installed via a Helm chart – `https://github.com/external-secrets/external-secrets/tree/main/deploy/charts/external-secrets` – and here is some very good documentation – `https://external-secrets.io/`. These are some great sources to read in order to get started.

Removing Dex Deployment

When we are using OIDC directly with Argo CD, or we're using no SSO at all, we can remove the Dex installation as it will not be useful for us. One way to do that is to set it to 0 Pods, but it is also possible with a Kustomize patch to remove the Deployment completely, and this is what we will do here. We need to make two changes – one is to add a new file called `argocd-dex-server.yaml`, put it in the `patches` folder, and that's where we will place the `delete` patch:

```
apiVersion: apps/v1
kind: Deployment
metadata:
 name: argocd-dex-server
$patch: delete
```

And the second is to add the entry in the `kustomization.yaml` file for the new files we added (showing here just the `patchesStrategicMerge` entries):

```
patchesStrategicMerge:
  - patches/argocd-cm.yaml
  - patches/argocd-rbac-cm.yaml
  - patches/argocd-dex-server-deployment.yaml
```

Next, create the commit and push it to remote so Argo CD can apply it. The Dex deployment will be removed and if we want to bring it back later, we need to remove this Kustomization patch file.

It is important to secure these OIDC connectivities as, most likely, your company will have a rule that all systems you use should have SSO. So, it would be a good practice for all sensitive information on SSO be kept secure, as it is not just about Argo CD anymore. Stopping Dex can also be seen as an improvement as we reduce the attack surface by stopping a service that we will not be using; for example, it will not affect us in any way if it will have any vulnerabilities in the future.

Summary

In this chapter, we went through different options for how we can accomplish user administration and access control. We started with how we can create new local users and assign them rights in order to be able to disable the admin account as it would be too risky to use it for daily operations. Then, we discovered what the options are for service accounts and how to access Argo CD from a pipeline or other types of automations. Local users are great for small companies, but SSO is becoming more relevant because it is more secure and has the power to manage everything from one place Argo CD supports many SSO providers either directly or via Dex; the official documentation has a lot of examples on how to set that up and we saw how we can make them more secure.

In the next chapter, we will dive deep into how to create a Kubernetes cluster and then deploy applications on it with Argo CD following the app-of-apps pattern.

Further reading

- Another example of how Attribute mapping can be used for Group mapping on Google SAML: https://www.dynatrace.com/support/help/how-to-use-dynatrace/user-management-and-sso/manage-users-and-groups-with-saml/saml-google-workspace#preparing-group-mapping

- Why Argo CD implemented the direct OIDC option and allows you to skip Dex: https://github.com/argoproj/argo-cd/issues/671.

- Benefits of SSO: https://www.onelogin.com/learn/why-sso-important.

- OpenID Connect protocol: https://openid.net/connect/faq/.

Part 3: Argo CD in Production

This part will cover the practical ways to understand and use Argo CD in production in a repeatable manner.

This part of the book comprises the following chapters:

- *Chapter 5, Argo CD Bootstrap K8s Cluster*
- *Chapter 6, Designing Argo CD Delivery Pipelines*
- *Chapter 7, Troubleshooting Argo CD*
- *Chapter 8, YAML and Kubernetes Manifests (Parsing and Verification)*
- *Chapter 9, Future and Conclusion*

5

Argo CD Bootstrap K8s Cluster

In this chapter, we will see how we can bootstrap K8s clusters in a repeatable automated manner with the necessary services and utilities ready for usage, and for disaster recovery purposes. We will go through the creation of a K8s cluster in AWS using **Infrastructure as Code** (**IaC**) tools, and in the post-creation phase, we will set up a K8s cluster with the required services and utilities using Argo CD. Then, we will identify how we can tackle the problem of dependencies in a cluster and how we can control their creation order with sync waves, which we discussed in *Chapter 2, Getting Started with Argo CD*.

At the end of the chapter, we will identify the security challenges of deploying services and following GitOps practices with Argo CD and how we can tackle them securely.

The main topics we will cover are as follows:

- Amazon EKS with Terraform
- Bootstrapping EKS with Argo CD
- Using the app of apps pattern
- Bootstrapping in practice
- ApplicationSet – the evolution

Technical requirements

For this chapter, we assume that you have already installed the Helm CLI. Additionally, you will need the following:

- An AWS account: https://aws.amazon.com/free.
- Terraform: https://learn.hashicorp.com/collections/terraform/aws-get-started.

- The AWS CLI: `https://docs.aws.amazon.com/cli/latest/userguide/getting-started-install.html`.

- The AWS IAM authenticator: `https://docs.aws.amazon.com/eks/latest/userguide/install-aws-iam-authenticator.html`.

- A code editor with YAML and Terraform support. I am using Visual Studio Code: `https://code.visualstudio.com`.

We are going to use the free AWS account to deploy with Terraform a managed K8s cluster, which is called Amazon EKS: *https://aws.amazon.com/eks/*. We will bootstrap EKS with the necessary services using Argo CD.

The code for this chapter can be found at `https://github.com/PacktPublishing/ArgoCD-in-Practice` in the `ch05` folder.

Amazon EKS with Terraform

In this section, we will create a managed K8s cluster in AWS so that we can use it in a real-time scenario to bootstrap ready-to-use K8s clusters in production environments. We will describe Amazon EKS, which is the managed K8s cluster of AWS, and how we can provision it with IaC and, more specifically, with Terraform.

Getting familiar with Amazon EKS

Most of the cloud providers have implemented managed K8s and they offer a fully managed control plane. AWS has Amazon EKS, which provides a fully managed and highly available control plane (K8s API server nodes and an etcd cluster).

Amazon EKS helps us to operate and maintain K8s clusters in AWS without a hassle. The API server, scheduler, and `kube-controller-manager` run in a VPC (`https://aws.github.io/aws-eks-best-practices/reliability/docs/controlplane/#eks-architecture`) managed by AWS in an auto-scaling group and different Availability Zones in an AWS Region. The following architecture diagram gives you a nice overview of Amazon EKS:

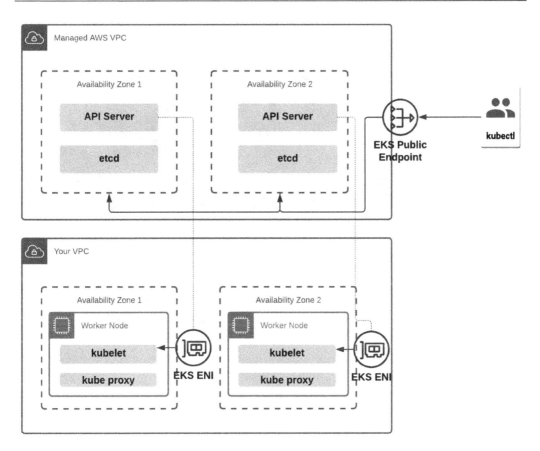

Figure 5.1 – Amazon EKS architecture

In practice, we only manage the worker node groups by ourselves, where we will deploy our services and which are included in our VPC, as described in *Figure 5.1*.

Let's now write the Terraform scripts that we will use to provision an Amazon EKS cluster in practice.

Designing EKS infrastructure

The first thing we need to do is to sign up for a free account in AWS so that we can deploy our infrastructure. You can find more details about how to set up an AWS account at this link: https://docs.aws.amazon.com/cli/latest/userguide/getting-started-prereqs.html#getting-started-prereqs-signup.

In order to interact with AWS from our local machine instead of the web console, we need to create an IAM user with the required permissions and generate the AWS credentials to configure the local AWS CLI.

In order to create an IAM user, we need to visit the IAM page in the AWS menu. Type I am in the top search bar; you can see where to find it in *Figure 5.2*:

Figure 5.2 – IAM management in AWS

Then, you can click on **IAM user** to create a user other than the AWS root account we used to sign up for a free AWS account. Check the following options in *Figure 5.3* to fill in and create an IAM user:

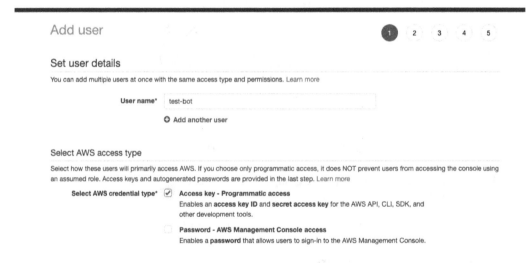

Figure 5.3 – Creating an IAM user

On the next screen, create a group for testing purposes with Administrator access by creating an admin policy name – in this case, TestGroup, as shown in *Figure 5.4*:

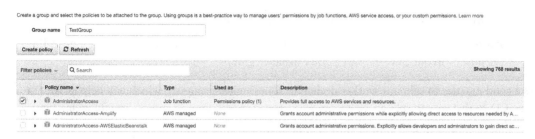

Figure 5.4 – Assigning an IAM group to an IAM user

Now, generate the AWS security credentials for programmatic access, as shown in the following screenshot:

Figure 5.5 – Assigning an IAM group to an IAM user

You need to download the CSV or copy the credentials to a secure place, such as a password manager. Now, it's time to configure the local AWS CLI so that we can interact with the AWS APIs. To do that, we need to run the following commands:

```
$ aws configure
AWS Access Key ID [None]: YOUR_AWS_ACCESS_KEY_ID
AWS Secret Access Key [None]: YOUR_AWS_SECRET_ACCESS_KEY
Default region name [None]: YOUR_AWS_REGION
Default output format [None]: json
```

The key infrastructure resources we need to create to provision an Amazon EKS cluster in the AWS cloud are the following:

- **VPC**: A virtual network where we can deploy AWS infrastructure resources.

- **Network Address Translation (NAT) gateway**: A NAT service that allows the EKS nodes in a private subnet to connect to services outside our VPC but does not allow any other service to initiate a connection with those nodes.

- **Security group**: A set of rules or, even better, a virtual firewall to control inbound and outbound traffic to our worker node groups.

- **Elastic Compute Cloud (EC2) instance**: A virtual server in AWS EC2.
- **Auto Scaling group**: A group of EC2 instances conform to a logical group for autoscaling purposes – in our case, the K8s worker instance groups.

It's a good practice to split our worker node into some logical grouping, which will help us later in Argo CD so that we can split and have separate groups for different purposes. The following figure describes this logical separation into two groups:

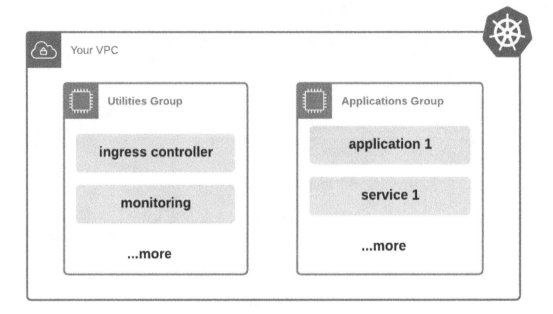

Figure 5.6 – Worker nodes' logical grouping

Let's define the purpose of each worker node group so that we can have a bit more clarity:

- **Utilities**: This group will be responsible for including utilities/tools to make the applications/ services available to internal and external users.
- **Application**: This group will be responsible for deploying any applications and services available to internal and external users.

Next, we are going to use Terraform, which is an OSS IaC tool created by HashiCorp. It gives us the ability to describe with a declarative configuration the resources of the infrastructure of a data center.

Provisioning EKS with Terraform

As we mentioned earlier, we need to create some necessary resources before we create an EKS cluster. The structure we are going to follow will be divided into logical groups of resources:

- `provider.tf`: The provider in Terraform is an abstraction of the AWS API.
- `network.tf`: The network layer where the resources will be created.
- `kubernetes.tf`: The EKS infrastructure resources and worker node groups.
- `variables.tf`: The dynamic variables, which we pass to Terraform scripts so that they can be repeatable.
- `outputs.tf`: The output of different resources, which can be printed out in the system console.

First, let's set up the related VPC where the EKS worker nodes groups will be created. Now, we need to define in Terraform the cloud provider we want to use in `provider.tf`:

```
provider "aws" {
  region = var.region
}
```

In Terraform, the community has built a couple of modules that abstract a set of resources that are used and created together. So, to create a new network layer (VPC) in our case, we will use a module provided by the AWS team in `network.tf`:

```
data "aws_availability_zones" "zones" {}
module "vpc" {
  source  = "Terraform-aws-modules/vpc/aws"
  version = "3.11.0"

  name                = var.network_name
  cidr                = var.cidr
  azs                 = data.aws_availability_zones.zones.
names
  private_subnets     = var.private_subnets
  public_subnets      = var.public_subnets
  enable_nat_gateway  = true
  enable_dns_hostnames = true

  tags = {
    "kubernetes.io/cluster/${var.cluster_name}" = "shared"
```

```
  }

  public_subnet_tags = {
    "kubernetes.io/cluster/${var.cluster_name}" = "shared"
    "kubernetes.io/role/nlb"                    = "1"
  }

  private_subnet_tags = {
    "kubernetes.io/cluster/${var.cluster_name}" = "shared"
    "kubernetes.io/role/internal-nlb"           = "1"
  }
}
```

The reason we used a new VPC is that, in any case, if your AWS account is not new and you already had something deployed inside it, to not impact existing resources and create the new EKS cluster there. Let's look at *Figure 5.7* to understand the network topology:

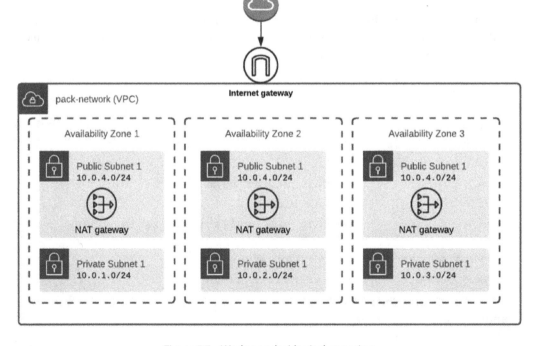

Figure 5.7 – Worker nodes' logical grouping

As we can see in *Figure 5.7*, we have a public and a private subnet for each Availability Zone and a NAT gateway for each public subnet. Private subnets can download packages and have access to the outside world, as they have attached the NAT gateway in their route table as a target. In *Figure 5.8*, we can see an example of the route table of a private subnet:

Main route table

Destination	Target
10.0.0.0/16	local
0.0.0.0/16	nat-gateway-id

Figure 5.8 – Worker nodes' logical grouping

Now, it's time to define the EKS K8s cluster, as described in *Figure 5.6*, and attach the private subnets to the K8s nodes so that they are not directly accessible from the outside world of the network. So, if we evolve *Figure 5.7* now with EKS K8s nodes, we will have the topology shown in *Figure 5.9*:

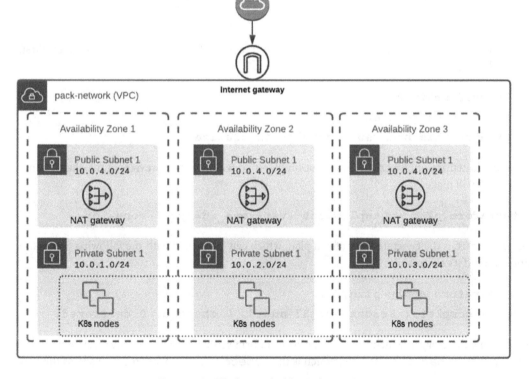

Figure 5.9 – Worker nodes' logical grouping

It is a good practice to always set the version of the providers we use in order to avoid a break in compatibility if there is a major version upgrade. We have a specific file to define these versions, which is versions.tf:

```
terraform {
  required_providers {
    aws = {
      source  = "hashicorp/aws"
      version = ">= 3.66.0"
    }

    kubernetes = {
      source  = "hashicorp/kubernetes"
      version = ">= 2.6.1"
    }
  }
}
```

Now, we are ready to run the Terraform scripts and create the AWS infrastructure resources. First, we need to initialize the Terraform working directory:

```
$ terraform init
Initializing modules...
Terraform has been successfully initialized!
```

After the initialization, we create an execution plan that will give us a preview of the changes that Terraform will make:

```
terraform plan -out=plan.out -var=zone_id=<your-zone-id>
```

In plan.out is the generated Terraform plan, which can be applied with the following command and output if it's successful:

```
$ terraform apply plan.out
Apply complete! Resources: 57 added, 0 changed, 0 destroyed.
```

In the same directory in which we did a run of Terraform scripts, there is a newly created file called kubeconfig_packt-cluster, which is the kubeconfig file we can use to access the new

EKS K8s cluster in AWS through the `kubectl` commands. Now, it's time to configure our terminal with this K8s configuration:

```
$ export KUBECONFIG=kubeconfig_packt-cluster
$ kubectl -n kube-system get po
```

The last command will return all the Pods that are under the `kube-system` namespace, and the output will be like that shown in *Figure 5.10*:

NAME	READY	STATUS	RESTARTS	AGE
aws-node-4tdgf	1/1	Running	0	14m
aws-node-bsjzp	1/1	Running	0	14m
aws-node-d554x	1/1	Running	0	14m
aws-node-jm25x	1/1	Running	0	14m
coredns-65bfc5645f-84wbt	1/1	Running	0	28m
coredns-65bfc5645f-nqx26	1/1	Running	0	28m
kube-proxy-87hbc	1/1	Running	0	14m
kube-proxy-cf7j9	1/1	Running	0	14m
kube-proxy-h7nlw	1/1	Running	0	14m
kube-proxy-ksrhn	1/1	Running	0	14m

Figure 5.10 – Pods available in kube-system

We now have an EKS K8s cluster ready in AWS, and we can jump into the hands-on work to bootstrap the cluster for repeatability and disaster-recovery scenarios. In the next chapter, we will have a post-creation phase of the initial infrastructure, where we will set Argo CD in the cluster along with the required utilities and applications.

Bootstrapping EKS with Argo CD

Now that we have created the cluster, we need to evolve our Terraform scripts a bit more so that for every new cluster we create, we set it up to be ready with Argo CD and manage the configuration so that we can follow GitOps practices for Argo CD too. As we mentioned in *Chapter 2, Getting Started with ArgoCD*, Argo CD syncs and manages itself.

Preparing Argo CD Terraform

We will use the Terraform provider for `kustomize` and install Argo CD in the cluster. First, we will create `kustomization.yaml`, which will install Argo CD and, in parallel, create the K8s resources for Argo CD, such as the name and the related `ConfigMap`. Here is `kustomization.yaml`:

```
apiVersion: kustomize.config.k8s.io/v1beta1
kind: Kustomization
```

```
namespace: argocd
bases:
  - https://raw.githubusercontent.com/argoproj/argo-cd/v2.1.7/
manifests/install.yaml
resources:
  - namespace.yaml
patchesStrategicMerge:
  - argocd-cm.yaml
```

With the preceding kustomization in practice, we will install Argo CD in a namespace resource, which is defined in a namespace.yaml file called argocd, and with a strategic merge (*https://github.com/ kubernetes/community/blob/master/contributors/devel/sig-api-machinery/strategic-merge-patch.md*), we add the ConfigMap for Argo CD too. Now, it's time to define in Terraform how to install Argo CD with kustomize in the new EKS cluster. First, we need to add an extra provider for kustomization and add the following in provider.tf:

```
provider "kustomization" {
  kubeconfig_path = "./${module.eks.kubeconfig_filename}"
}
```

Here is argocd.tf to bootstrap the cluster with Argo CD:

```
data "kustomization_build" "argocd" {
  path = "../k8s-bootstrap/bootstrap"
}
resource "kustomization_resource" "argocd" {
  for_each = data.kustomization_build.argocd.ids
  manifest = data.kustomization_build.argocd.manifests[each.
value]
}
```

Now, it's time to apply the Terraform script to install Argo CD.

Applying Argo CD Terraform

Similar to what we did when we created the EKS K8s cluster, we need to plan the changes in Terraform:

```
$ terraform plan -out=plan.out
```

This command will display that there are a couple of changes:

```
Plan: 41 to add, 0 to change, 0 to destroy.
```

Now is the time to apply and check that everything works as expected:

```
$ terraform apply plan.out
```

In the end, everything will be successful, and you will see this message:

```
Apply complete! Resources: 20 added, 0 changed, 0 destroyed.
```

But is Argo CD really installed and running? Let's check it with the following command:

```
$ kubectl -n argocd get all
```

Everything worked, and you should be able to see something similar to *Figure 5.11*:

```
NAME                                        READY   STATUS    RESTARTS   AGE
pod/argocd-application-controller-0         1/1     Running   0          86s
pod/argocd-repo-server-6fd99dbbb5-rswbc     1/1     Running   0          86s
pod/argocd-server-7674b5cff5-szj77          1/1     Running   0          88s

NAME                            TYPE        CLUSTER-IP       EXTERNAL-IP   PORT(S)                       AGE
service/argocd-dex-server       ClusterIP   172.20.226.115   <none>        5556/TCP,5557/TCP,5558/TCP    4m17s
service/argocd-metrics          ClusterIP   172.20.61.41     <none>        8082/TCP                      4m15s
service/argocd-redis            ClusterIP   172.20.28.95     <none>        6379/TCP                      90s
service/argocd-repo-server      ClusterIP   172.20.22.22     <none>        8081/TCP,8084/TCP             90s
service/argocd-server           ClusterIP   172.20.151.64    <none>        80/TCP,443/TCP                90s
service/argocd-server-metrics   ClusterIP   172.20.246.236   <none>        8083/TCP                      4m15s

NAME                                  READY   UP-TO-DATE   AVAILABLE   AGE
deployment.apps/argocd-dex-server     0/1     0            0           4m18s
deployment.apps/argocd-redis          0/1     0            0           4m18s
deployment.apps/argocd-repo-server    1/1     1            1           88s
deployment.apps/argocd-server         1/1     1            1           91s

NAME                                             DESIRED   CURRENT   READY   AGE
replicaset.apps/argocd-dex-server-5896d988bb     1         0         0       4m19s
replicaset.apps/argocd-redis-74d8c6db65          1         0         0       4m19s
replicaset.apps/argocd-repo-server-6fd99dbbb5    1         1         1       89s
replicaset.apps/argocd-server-7674b5cff5         1         1         1       92s

NAME                                                READY   AGE
statefulset.apps/argocd-application-controller      1/1     90s
```

Figure 5.11 – Bootstrapping EKS with Argo CD

Let's run a port-forward to the Argo CD K8s service and check whether the UI works, and we can see the argo-cd app managing itself as an Argo application. Let's fetch the password for the UI login first:

```
$ kubectl -n argocd get secret argocd-initial-admin-secret -o
jsonpath="{.data.password}" | base64 -d
```

It's not a good practice to use the admin user, but for convenience, we will use it. So, the username is admin, and the admin password is fetched by the preceding command. Let's now port-forward to access the UI with the following command:

```
$ kubectl -n argocd port-forward service/argocd-server 8080:80
```

We can see in *Figure 5.12* that the argo-cd application has been created and is already synced:

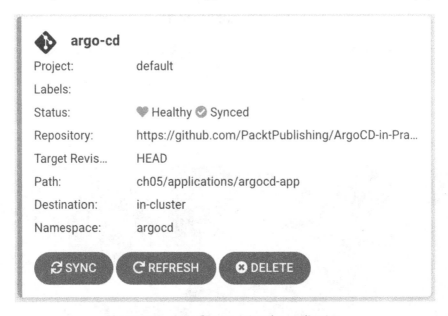

Figure 5.12 – Argo CD managing the application

If we check a bit further, we can see that argo-cm.yaml is already synced:

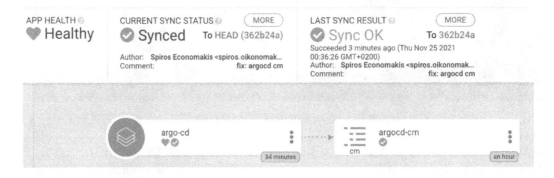

Figure 5.13 – Argo CD managing the application

In *Figure 5.13*, you can see that the app is healthy and synced without any issue. From this moment, we are able to change the configuration of Argo CD with GitOps, and in practice, as we mentioned earlier, Argo CD syncs and manages itself.

As a result, we have a new EKS cluster with Argo CD bootstrapped, and we have a repeatable manner with Terraform to create a new K8s cluster ready for use. In the next section, we will go a bit deeper into understanding the app of apps pattern and try to orchestrate the dependencies between the different utilities and applications we want to have bootstrapped in the cluster.

Using the app of apps pattern

In *Chapter 2, Getting Started with Argo CD*, we created a quick example of the app of apps pattern while we were using Argo CD Autopilot. It's also important to talk about how we would use the app of apps pattern in real environments and why we need it. In the previous section, we followed one step of the pattern by bootstrapping the cluster with Argo CD, but in a real environment, we want to create a repeatable K8s cluster with the same services; this is especially critical in disaster-recovery scenarios.

Why the app of apps pattern?

The pattern in practice gives us the ability to have a master application and logical grouping of another set of applications. You can create a master app that creates other apps and allows you to declaratively manage a group of apps that can be deployed in concert.

The pattern is used for bootstrapping a cluster with all the applications, and this is the main thing we are trying to accomplish in this chapter. The first issue, although simple to tackle, is how we can deploy our services and applications with GitOps practices. Another issue that an Argo CD application may have is that it can have only one source type, which means you cannot have Helm, kustomization, or plain YAML manifests in parallel.

With the app of apps pattern, instead of deploying multiple Argo CD application CRDs, you can deploy only one – the master app, which will deploy the rest for you. In the real world, even at the company that I am working for right now, we do logical grouping, which gives you the ability to have a watcher application that ensures all applications in the group are deployed and healthy.

Let's jump back into our bootstrapping of the EKS cluster and start adding the utilities and applications for our scenario.

Bootstrapping utilities

In the previous section, we created the Argo CD, and in order to access the UI, we ran a port-forward to the service of K8s, and then we were able to access it. In the real world, we have teams that need external access to utilities or services other than port-forward. The first thing we will try here is to create an Ingress and a way to give a DNS name so that we can access them accordingly when needed.

For the sake of the use case here, we will use just public access in AWS and the public hosted zones, although, in a real environment, we always maintain one public and one private hosted zone. The reason is to give access to private hosted zones only through a VPN for services that are not for public access.

We need two utilities in order to have an Ingress and a DNS name. Specifically, we are going to use the following:

- **External DNS**: This will help create AWS Route 53 records for the Istio Gateway ingresses.
- **Istio operator**: This will help install, upgrade, and uninstall Istio in an environment, and will also install Istio and a control plane.
- **Istio control plane**: This will help to manage and configure the ingress gateways to route traffic.
- **Istio resources**: This will help you to access the Argo CD UI without port-forwarding.

The important part is the order of deploying these utilities in order to have them up and running and the logical grouping of the app of apps pattern. The order we need to follow is the following:

1. External DNS
2. The Istio operator
3. The Istio control plane
4. Argo CD Istio ingress

I think you already know the answer after reading *Chapter 2, Getting Started with Argo CD*, to how we are going to achieve the ordering. Of course, the answer is `sync-waves`, so let's now see them in practice. First, we need to define the External DNS in the `master-utilities` Argo application. Before we start with the YAML files, you need to know that we need a registered domain name so that we can use it as an AWS public hosted zone in Route 53 (`https://aws.amazon.com/route53/`). Then, the External DNS will use the public hosted zone to create the related records defined in Istio as the Gateway.

AWS offers the ability to register domain names and automatically create an AWS Route 53 hosted zone after registration has been approved. In this case, I am using the registered domain name and hosted zone at `http://packtargocdbook.link/`, as shown in *Figure 5.14*:

Figure 5.14 – Argo CD managing the application

Next, we need to create an IAM role with a specific IAM policy to list and change resource record sets in AWS Route 53. The following is the related Terraform script to create the necessary role in `iam.tf` with the following policy:

```
resource "aws_iam_policy" "external_dns" {
  name        = "external-dns-policy"
  path        = "/"
  description = "Allows access to resources needed to run
external-dns."
  policy = <<JSON
{
  "Version": "2012-10-17",
  "Statement": [
    {
      "Effect": "Allow",
      "Action": [
        "route53:ChangeResourceRecordSets"
      ],
      "Resource": [
        "${data.aws_route53_zone.zone_selected.arn}"
      ]
    },
    {
      "Effect": "Allow",
      "Action": [
        "route53:ListHostedZones",
        "route53:ListResourceRecordSets"
      ],
      "Resource": [
        "*"
      ]
    }
  ]
}
JSON
}
```

The first application is the External DNS one and you can find it in the official Git repository: `https://github.com/PacktPublishing/ArgoCD-in-Practice` in the `ch05/applications/master-utilities/templates/external-dns.yaml` file.

Next in order is the Istio operator in `applications/master-utilities/istio-operator.yaml`:

```yaml
apiVersion: argoproj.io/v1alpha1
kind: Application
metadata:
  name: istio-operator
  namespace: argocd
  annotations:
    argocd.argoproj.io/sync-wave: "-2"
  finalizers:
    - resources-finalizer.argocd.argoproj.io
spec:
  project: default
  source:
    repoURL: https://github.com/istio/istio.git
    targetRevision: "1.0.0"
    path: operator/manifests/charts/istio-operator
    helm:
      parameters:
        - name: "tag"
          value: "1.12.0"
  destination:
    namespace: istio-operator
    server: {{ .Values.spec.destination.server }}
  syncPolicy:
    automated:
      prune: true
```

Next is the Istio control plane, and here is the Argo app – `applications/master-utilities/istio.yaml`:

```yaml
apiVersion: argoproj.io/v1alpha1
kind: Application
metadata:
```

```
  name: istio
  namespace: argocd
  annotations:
    argocd.argoproj.io/sync-wave: "-1"
  finalizers:
    - resources-finalizer.argocd.argoproj.io
spec:
  project: default
  source:
    repoURL: https://github.com/PacktPublishing/ArgoCD-in-
Practice.git
    targetRevision: HEAD
    path: ch05/applications/istio-control-plane
  destination:
    namespace: istio-system
    server: {{ .Values.spec.destination.server }}

  syncPolicy:
    automated:
      prune: true
    syncOptions:
      - CreateNamespace=true
```

Last but not the least, we need the Argo application with the Istio resource for Argo CD UI access:

```
apiVersion: argoproj.io/v1alpha1
kind: Application
metadata:
  name: argocd-istio-app
  namespace: argocd
  finalizers:
    - resources-finalizer.argocd.argoproj.io
spec:
  project: default
  source:
    repoURL: https://github.com/PacktPublishing/ArgoCD-in-
Practice.git
    targetRevision: HEAD
```

```
        path: ch05/applications/argocd-ui
    destination:
      namespace: argocd-ui
      server: {{ .Values.spec.destination.server }}
    syncPolicy:
      automated:
        prune: true
        selfHeal: true
      syncOptions:
        - CreateNamespace=true
      validate: true
```

However, let's mention something important here. The order is defined by the sync waves that are set as Argo CD annotations. We will see in the Manifest file that the Istio operator chart has been assigned `argocd.argoproj.io/sync-wave: "-2"` and, conversely, the Istio control plane chart has `argocd.argoproj.io/sync-wave: "-1"`. The end result is that the Istio operator will be installed first, as it has a lower value than the Istio control plane. Argo CD will wait to finish the deployment and be in a healthy state in order to proceed to the next wave.

There is a small problem though, which is that Argo CD doesn't really know when the Istio control plane is healthy, meaning that it doesn't know when the Istio resource has been created. The only thing that Argo CD knows is that it's deployed and healthy when the chart is installed in the EKS cluster.

The preceding problem can be solved with the custom health checks by Argo CD, which you can write in a programming language called Lua. There are two ways we can define custom health checks:

- Use the argocd-cm.yaml `ConfigMap` so that we can leverage GitOps too
- Custom health checks scripts as part of an `argo-cd` build, which are not very flexible

We are going to use the first option, as it's more flexible, and we need to add a custom health check with a Lua script. The easy part here is that the Istio operator provides a status attribute in the `IstioOperator` CRD, which will be updated once the Istio control plane resources have been created successfully. We will modify `argocd-cm.yaml` to the following:

```
apiVersion: v1
kind: ConfigMap
metadata:
  name: argocd-cm
  namespace: argocd
  labels:
    app.kubernetes.io/name: argocd-cm
```

```
      app.kubernetes.io/part-of: argocd
data:
  resource.customizations: |
    install.istio.io/IstioOperator:
      health.lua: |
        hs = {}
        if obj.status ~= nil then
          if obj.status.status == "HEALTHY" then
            hs.status = "Healthy"
            hs.message = "Istio-Operator Ready"
            return hs
          end
        end

        hs.status = "Progressing"
        hs.message = "Waiting for Istio-Operator"
        return hs
```

The Lua script added here will be used periodically by Argo CD to validate the health status of K8s objects of the `install.istio.io/IstioOperator` type. Now, after the modification, we can just commit and push the change because Argo CD manages itself, and if there is a drift of changes, it will apply them. Easy, right?

In the next section, we will start from the beginning, and we will first destroy everything completely so that we can see in practice the bootstrap of a new cluster ready for usage.

Bootstrapping in practice

In this section, the end goal is to create a fresh EKS cluster and validate all the actions we made in the previous sections. The validation criteria are the following:

- The EKS cluster created
- Two worker node groups, one for utilities and one for applications
- Accessing the UI of Argo CD (meaning the Istio operator and master-utilities are deployed already as they were part of previous sync-waves)
- Checking whether the Argo CD application to manage itself is there

Now, it's time to destroy the infrastructure and recreate the infrastructure so that we can cross-validate what we have done so far is working, as described in the previous sections.

Destroying infrastructure

We have bootstrapped the EKS cluster with the necessary utilities in this chapter, and now it's time to destroy the infrastructure completely and start afresh. Terraform gives us the ability to destroy infra with the following command to completely clean up the AWS account of all the resources:

```
$ terraform destroy -auto-approve
```

Starting fresh infrastructure

Once the infrastructure is destroyed, it's time to recreate everything from the beginning. The first thing we need to do is to clean up the Terraform workspace:

```
$ rm -rf .terraform .terraform.lock.hcl *.tfstate*
```

We need to initialize Terraform so that we can redownload the providers we have set with the related versions:

```
$ terraform init
...·····
Terraform has been successfully initialized!
```

Now, it is time to first plan the Terraform script and then apply it. After some time – it will take a couple of minutes – the apply will finish successfully, and here are the related commands:

```
$ terraform plan -out=plan.out
$ terraform apply plan.out -auto-approve
```

After everything is done, we are ready to validate whether the expected result and all the utilities have been installed and are in a healthy state in Argo CD.

In the previous section, we discussed and deployed an app of apps pattern for the sake of our example. The Argo team evolved the app of apps pattern and created a new Argo CD CRD called ApplicationSet. In the next section, we will try to understand some of the disadvantages of the app of apps pattern and how ApplicationSet tackled them.

App of apps disadvantages

The pattern solved a couple of the problems of bootstrapping a cluster easily and helped us to avoid deploying hundreds of Argo CD application objects, as you actually just need to deploy one that groups all the others. A big plus was that this pattern gave us the ability to have an observer Argo CD application that ensures that the rest of the applications are deployed and healthy.

However, we still have some challenges to solve, such as how to support *monorepos* or *multi-repos*, especially in a microservice world, how to give a team the ability to deploy applications using Argo CD in multi-tenant clusters without the need for privilege escalation, and, of course, how to avoid creating too many apps.

Enter the newest controller by the Argo team, `ApplicationSet`, which supplements the Argo CD application functionality. We will see how it works in the next section.

What is ApplicationSet?

The `ApplicationSet` controller is a typical K8s controller and works along with Argo CD to manage Argo CD applications, and we can think of it as an application factory. The great features that the `ApplicationSet` controller gives us are the following:

- One K8s manifest to deploy to multiple K8s clusters with Argo CD

- One K8s manifest to deploy multiple Argo applications by multiple sources/repositories

- Great support for "*monorepos*", which in Argo CD means multiple Argo CD application resources in one repository

In *Figure 5.15*, we can see how the `ApplicationSet` controller interacts with Argo CD and that the only purpose it has is to create, update, and delete Argo application resources within the Argo CD namespace. The only goal is to ensure that Argo apps have the right state, which has been defined in the declarative `ApplicationSet` resource:

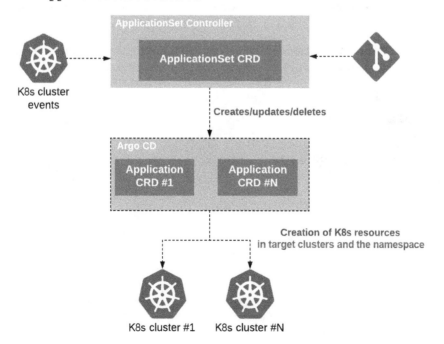

Figure 5.15 – The application's controller interaction with Argo CD

`ApplicationSet` relies on generators to generate parameters for the different data source supports. We have the following generators:

- **List**: A fixed list of clusters to target Argo CD applications.

- **Cluster**: A dynamic list based on the clusters is already defined and managed by Argo CD.

- **Git**: The ability to create applications within a Git repository or ones based on the directory structure of a Git repository.

- **Matrix**: Combines parameters for two different generators.

- **Merge**: Merges parameters for two different generators.

- **SCM provider**: Automatically discovers the repositories in an **Source Code Management (SCM)** provider (for example, GitLab and GitHub).

- **Pull request**: Automatically discovers open pull requests in an SCM provider (for example, GitLab and GitHub).

- **Cluster decision resource**: Generates a list of clusters in Argo CD.

It's time to evolve our current CI/CD a bit to a monorepo approach with multiple services and utilize the `ApplicationSet` controller.

Generators

The fundamental components of `ApplicationSet` are generators, which are responsible for generating parameters that are then used in the fields of `ApplicationSet`.

Let's look at some of the available generator types. First, we will look at the **List generator**, which will allow us to target Argo applications to a fixed list of clusters. An example can be found here:

```
apiVersion: argoproj.io/v1alpha1
kind: ApplicationSet
metadata:
  name: chaos-engineering
spec:
  generators:
  - list:
      elements:
      - cluster: cloud-dev
        url: https://1.2.3.4
      - cluster: cloud-prod
        url: https://2.4.6.8
```

```
      - cluster: cloud-staging
        url: https://9.8.7.6
    template:
      metadata:
        name: '{{cluster}}- chaos-engineering
      spec:
        source:
          repoURL: https://github.com/infra-team/chaos-
engineering.git
          targetRevision: HEAD
          path: chaos-engineering/{{cluster}}
        destination:
          server: '{{url}}'
          namespace: guestbook
```

Note that we have defined a list of clusters, such as cloud-dev, cloud-prod, and cloud-staging. We can make the preceding manifest a bit more flexible when we don't want to have a fixed list, and we can use ClusterGenerator for this. With ClusterGenerator, we have two options:

- Target all the K8s clusters available in Argo CD

- Target the K8s cluster that matches a label selector

Let's target all the clusters available in Argo CD first and see the difference with ListGenerator:

```
apiVersion: argoproj.io/v1alpha1
kind: ApplicationSet
metadata:
  name: chaos-engineering
spec:
  generators:
  - clusters: {}
  template:
    metadata:
      name: '{{cluster}}-chaos-engineering'
    spec:
      source:
        repoURL: https://github.com/infra-team/chaos-
engineering.git
        targetRevision: HEAD
```

```
        path: chaos-engineering/{{cluster}}
      destination:
        server: '{{url}}'
```

Actually, the only difference is the property clusters in `manifest: {}`, which means that if we don't define anything, we will do the application to every available cluster in Argo CD.

With a label selector, we need to have a metadata label in the Secret for the cluster, which we can use to select and deploy only to these clusters. An example is the following:

```
kind: Secret
data:
  # etc.
metadata:
  labels:
    argocd.argoproj.io/secret-type: cluster
    sre-only: "true"
```

Here is `ApplicationSet`, which selects the clusters by the matching label:

```
kind: ApplicationSet
metadata:
  name: chaos-engineering
spec:
  generators:
  - clusters:
      selector:
        matchLabels:
          sre-only: true
```

That's it! Now, it's clear that the app of apps pattern will soon be replaced by the powerful `ApplicationSet`, which can be very helpful when you have multiple utilities and services to deploy in concert.

Summary

In this chapter, we learned step by step how to create a new EKS cluster in AWS and bootstrapped it in an instrumented manner by covering Argo CD, learning how to use an Argo CD app to manage itself, External DNS, the Istio operator, the Istio control plane, and the Argo CD Istio ingress.

The full code can be found at `https://github.com/PacktPublishing/ArgoCD-in-Practice` in the `ch05` folder. We gained a better understanding of sync waves in a real scenario that we can apply to our day-to-day work operations. In the end, we learned about the new custom health checks that, through a simple Lua programming script, can give us the power to validate when a resource is really ready.

In the next chapter, we will see how we can adopt Argo CD in a CI/CD workflow, tackle the security challenges with Secrets, and coordinate a real scenario of microservice deployments with Argo Rollouts.

Further reading

- Cluster bootstrapping: `https://argo-cd.readthedocs.io/en/stable/operator-manual/cluster-bootstrapping/`

- GitOps and K8s bootstrapping: `https://medium.com/lensesio/gitops-and-k8s-bootstrapping-752a1d8d7085`

- Sync waves: `https://argo-cd.readthedocs.io/en/stable/user-guide/sync-waves/`

- Custom health checks: `https://argo-cd.readthedocs.io/en/stable/operator-manual/health/#custom-health-checks`

- `ApplicationSet`: `https://github.com/argoproj-labs/applicationset`

6
Designing Argo CD Delivery Pipelines

In this chapter, we will use the infrastructure we created in *Chapter 5, Argo CD Bootstrap K8s Cluster,* to demonstrate real deployment strategies using Argo CD and get familiarized with Argo Rollouts. We will define a hypothetical scenario of real engineering with multiple teams and microservices and write a real CI pipeline that interacts with Argo CD, leveraging the Argo CLI and Argo RBAC security for different Argo Projects.

At the end, we will try to tackle the challenges of GitOps, as everything is in a Git repository and we need to find a way to keep the secrets of our services safe.

The main topics we will cover are the following:

- Motivation
- Deployment strategies
- Keeping secrets safe
- Real CI/CD pipeline
- Microservices CI/CD in practice

Technical requirements

For this chapter, we assume that you have already installed the Helm CLI and you have already run the Terraform scripts from *Chapter 5, Argo CD Bootstrap K8s Cluster.* Additionally, you will need the following:

- Basic Golang
- GitHub Actions: `https://docs.github.com/en/actions`

- External Secrets: `https://github.com/external-secrets/external-secrets`
- `curl`: `https://curl.se/`

The code can be found at `https://github.com/PacktPublishing/ArgoCD-in-Practice` in the `ch05` and `ch06` folders.

Motivation

Some companies are already trying to transition their services from traditional VMs in the cloud or other container orchestration tools (for example, AWS ECS, Azure CI, and so on) to K8s clusters. One of the biggest problems though in moving to K8s is how we can set up more sophisticated deployment strategies, as the standard rolling updates provided for free by K8s would not work in some cases. For example, what if I want to deploy a new version of a service but I want to first test that it's functional and then switch to the new version and, in parallel, destroy the old one? This is called blue/green deployment and can give me the power to reduce downtime and not even impact my end users.

In the next section, we will see how we can achieve this in K8s using only the K8s objects and how we can deploy a service with blue/green deployment.

Simple blue-green in K8s

I have created a small Golang app that serves a simple HTTP server and returns the version of the application under `localhost:3000/version`. I am leveraging Go build tags so I can have two variants for the different versions and responses of the HTTP server. The main code base is as follows:

```
package main
import (
        "fmt"
        "net/http"
)


var BuildVersion = "test"

func main() {
        http.HandleFunc("/version", version)
        http.ListenAndServe(":3000", nil)
}
func version(w http.ResponseWriter, r *http.Request) {
        fmt.Fprintf(w, BuildVersion)
}
```

Then, we will use Golang `ldflags` to change the response of the endpoint based on the different versions we want to have:

```
TAG=v1.0 make build-linux # or build-mac for Mac users
TAG=v2.0 make build-linux # or build-mac for Mac users
```

You can find a `Makefile` in the repository folder I mentioned earlier, which will build a Docker image for each different version. You can do this with the following command:

```
TAG=v1.0 make build-image # For v1.0 version
TAG=v2.0 make build-image # For v2.0 version
```

For your convenience, I have already built the Docker images and pushed them into Docker Hub as public images.

Now it's time to experiment with blue/green. First, let's set up our environment again so we can have access to the K8s cluster:

```
export KUBECONFIG= <your-path-to-book-repo>/ArgoCD-in-Practice/
ch05/terraform/kubeconfig_packt-cluster
```

Now, let's create a separate namespace to run the K8s manifest for the v1.0 version (blue) of the service:

```
kubectl create ns ch06-blue-green
```

Then apply `blue.yaml` under `ch06/simple-blue-green/deployments` with the following command:

```
kubectl apply -f blue.yaml
```

After applying it, you should be able to successfully see two Pods running with the v1.0 service version. The output would be similar to this:

```
NAME                    READY   STATUS    RESTARTS   AGE
app-f6c66b898-2gwtz     1/1     Running   0          108s
app-f6c66b898-fg2fv     1/1     Running   0          108s
```

Right now, it's time to deploy the K8s service, which will expose the set of deployed Pods as a network service. Now is the time to apply `service.yaml` under `ch06/ simple-blue-green/ deployments` and then check under the newly created AWS load balancer to see that we deployed the right version of the service:

```
kubectl apply -f service-v1.yaml
```

After the apply, an EXTERNAL-IP will be generated and, more specifically, an AWS load balancer as the K8s service is set as LoadBalancer. In my case, the output Loadbalancer, EXTERNAL-IP, is an AWS **Elastic Load Balancer (ELB)** (https://aws.amazon.com/elasticloadbalancing/) like this: aa16c546b90ba4b7aa720b21d93787b8-1761555253.us-east-1.elb. amazonaws.com. Let's run an HTTP request under /version so we can validate the version of the service:

```
Request:
curl aa16c546b90ba4b7aa720b21d93787b8-1761555253.us-east-1.elb.
amazonaws.com:3000/version
Response:
v1.0
```

Now it's time to run the K8s manifest for the v2.0 version (green) of the service:

```
kubectl apply -f green.yaml
```

Now we have two extra Pods that run the v2.0 (green) version of the service. Let's see the output after running kubectl -n ch06-blue-green get po:

```
NAME                      READY   STATUS    RESTARTS   AGE
app-f6c66b898-2gwtz       1/1     Running   0          12h
app-f6c66b898-fg2fv       1/1     Running   0          12h
app-v2-6c4788bf64-ds4dj   1/1     Running   0          3s
app-v2-6c4788bf64-lqwvf   1/1     Running   0          3s
```

If we make a request again with the AWS ELB, we will still see v1.0 as the K8s service still matches the deployment of v1.0 because of the following matching labels:

```
selector:
  app:  app
  version: "1.0"
```

Now, let's look at the current state of our deployment, as shown in *Figure 6.1*:

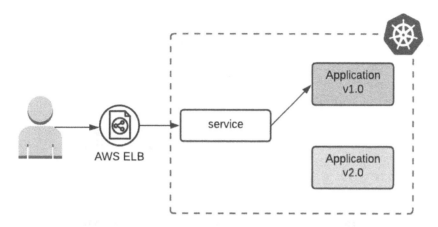

Figure 6.1 – Deployment serves blue version

Most engineering teams in this state will run automated or manual tests to validate that v2.0 is functional and won't cause any issues. Assuming that everything is functional, we can serve v2.0 of the service, and we need to change the selector labels in the K8s service to the following:

```
selector:
    app:   app
    version: "2.0"
```

After we have made the change, it's time to apply the service again:

```
kubectl apply -f service-v2.yaml
```

Let's run an HTTP request under /version so we can validate the version of the service:

```
Request:
curl aa16c546b90ba4b7aa720b21d93787b8-1761555253.us-east-1.elb.
amazonaws.com:3000/version
Response:
v2.0
```

Here, we have a new state diagram:

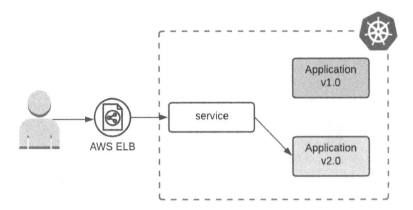

Figure 6.2 – Deployment serves green version, switching traffic

That's it! We just completed the first blue/green deployment in a K8s cluster. All the actions we took were manual, but we could automate these steps in a CI/CD pipeline, although that could be very complex and hard to maintain in the long run. Apart from this, we are not following GitOps principles, and we also need the K8s credentials exposed in CI/CD.

Here comes the Argo project that introduced Argo Rollouts, which is a K8s controller that will help us to make a progressive delivery of our services with more complex deployment strategies than the one we described here. Let's see in detail how Argo Rollouts work in the next section.

Deployment strategies

In this section, we will explain what Argo Rollouts is and take a deep dive into the architecture. We will also learn about the supported deployment strategies. At the end, we will run a real example of delivering microservices with a progressive delivery approach and recover automatically from failed deployments.

What is Argo Rollouts?

Argo Rollouts is a Kubernetes controller similar to the K8s Deployment objects but is a **Custom Resource Definition (CRD)** developed by the Argo project team. This CRD has extended capabilities so it can provide progressive delivery in Deployments such as the following:

- Blue-green deployments
- Canary deployments
- Weighted traffic shift
- Automatic rollbacks and promotions
- Metric analysis

Next, we will explain the reasons to use Argo Rollouts and the limitations of the default K8s rolling update strategy.

Why Argo Rollouts?

The standard K8s Deployment object only gives us the ability for the `RollingUpdate` strategy, which supports basic safety requirements during an update of a deployment such as a readiness probe. The rolling update strategy has many limitations though:

- There is no way to control how fast or slow we can make a rollout.

- Not able to have traffic switch in a newer version for cases such as canary or blue-green.

- There is no input by metrics that we can use for the rollout.

- Only the ability to halt the progression but there is no way to automatically abort and roll back an update.

For large-scale environments, the rolling update strategy is risky as it cannot provide control over the blast radius of a deployment. Next, we will see an overview of the Argo Rollouts architecture.

The architecture of Argo Rollouts

In practice, the Argo Rollouts controller is responsible for managing the life cycle of `ReplicaSets` as the typical K8s deployment object does. In *Figure 6.3*, we can see the architecture diagram for a canary deployment strategy, which we will explain in detail later:

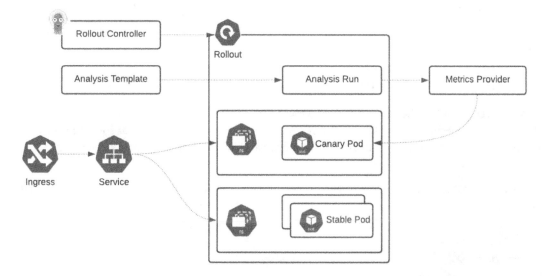

Figure 6.3 – Deployment serves green version, switching traffic

Let's see the purpose of each component in the architecture so we can understand how Argo Rollouts works:

- **Argo Rollouts controller**: As with every controller in K8s, this one also observes whether there are new object resources (custom resources) of a type called `Rollout`; the controller will then check the declaration of the Rollout and will try to bring the cluster to the declared state.

- **Rollouts resource (CRD)**: This is compatible with the typical K8s deployment object but it includes some extra fields that can control the stages and thresholds of the deployment strategies such as blue-green or canary. So, in practice, to leverage Argo Rollouts, you need to migrate your existing K8s deployments objects to the Rollouts resource object so they can be managed.

- **ReplicaSets**: This is the standard K8s `ReplicaSet` resource with extra metadata so the Argo Rollouts controller can track the different versions that are part of a deployment/application.

- **Analysis**: This is the intelligent part of Argo Rollouts that connects the Rollouts controller with our preferred metrics provider so we can define the metrics that would get a decision on whether the update completed successfully or not. If the metrics are validated and are good, then it will progress to deliver. On the other side, it will roll back if there is a failure or pause the rollout if the metrics provider cannot give an answer. This needs two K8s CRDs: **AnalysisTemplate** and **AnalysisRun**. **AnalysisTemplate** contains the details for which metric to query so it can get back a result, which is called **AnalysisRun**. The template can be defined in a specific rollout or globally on the cluster so it can be shared by multiple rollouts.

- **Metrics providers**: Integrations with tools such as Prometheus and Datadog that we can use in the `Analysis` component and do the clever part of automatically promoting or rolling back a rollout.

In the next section, we will use the blue-green deployment strategy supported by Argo Rollouts so that we can demonstrate how we can reduce the amount of time running multiple versions in parallel and, of course, deliver a stable newer version.

Blue-green deployment strategy

As we described earlier, the blue-green deployment has two versions of the application in parallel at the same time, specifically the old one and the new one. The production traffic for some time flows to the old version until the test suite runs (manually or automated) against the new version and then switches the traffic to the latest version.

We can achieve this with the Argo Rollouts controller. An example of the blue-green `Rollout` CRD is as follows:

```
kind: Rollout
metadata:
  name: rollout-bluegreen
spec:
```

```
replicas: 2
revisionHistoryLimit: 2
selector:
  matchLabels:
    app: bluegreen
template:
  metadata:
    labels:
      app: bluegreen
  spec:
    containers:
    - name: bluegreen-demo
      image: spirosoik/cho06:v1.0
      imagePullPolicy: Always
      ports:
      - containerPort: 8080
strategy:
  blueGreen:
    activeService: bluegreen-v1
    previewService: bluegreen-v2
    autoPromotionEnabled: false
```

The key difference from the typical K8s Deployment resource is the strategy where we define the deployment type and the two services that are used for the blue-green versions and that, in parallel, the Argo Rollouts controller will be triggered when we change the version of the image set in `template.spec`. The important fields in `blueGreen` are the following:

- `activeService`: This is the current blue version that will be served until the time we will run the promotion.

- `previewService`: This is the new green version that will be served after we promote it.

- `autoPromotionEnabled`: This decides whether the green version will be promoted immediately after `ReplicaSet` is completely ready and available. In the next section, we will describe what a canary deployment is and how we can define this in Argo Rollouts.

Canary deployment strategy

The idea behind the canary release is that we can serve a version/deployment to only a subset of end users while we serve the rest of the traffic to the old version. With the canary release, we can validate in reality whether the new version works correctly, and we can gradually increase the adoption of the end users and completely replace the old version. An example of a canary release is described in *Figure 6.4*:

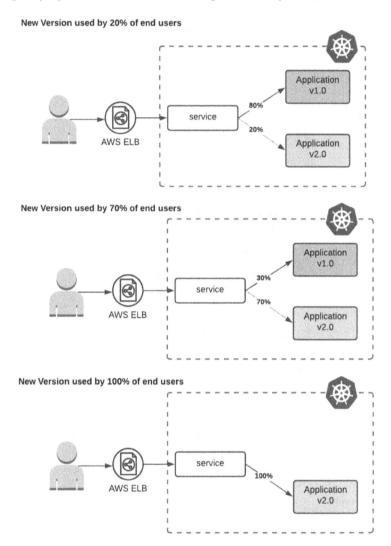

Figure 6.4 – Canary release

We can achieve this with the Argo Rollouts controller. An example of the canary `Rollout` CRD is as follows:

```
apiVersion: argoproj.io/v1alpha1
kind: Rollout
```

```
metadata:
  name: rollout-canary
spec:
  replicas: 5
  revisionHistoryLimit: 2
  selector:
    matchLabels:
      app: rollout-canary
  template:
    metadata:
      labels:
        app: rollout-canary
    spec:
      containers:
      - name: rollouts-demo
        image: spirosoik/ch06:blue
        imagePullPolicy: Always
        ports:
        - containerPort: 8080
  strategy:
    canary:
      steps:
      - setWeight: 20
      - pause: {}
      - setWeight: 40
      - pause: {duration: 40s}
      - setWeight: 60
      - pause: {duration: 20s}
      - setWeight: 80
      - pause: {duration: 20s}
```

The main difference here is the `canary` section; let's describe the steps for canary definition. The current rollout will start with a canary weight of 20% and as you can see the pause is set to { }, which means the rollout will pause indefinitely. We can explicitly resume the rollout with the following:

```
$ kubectl argo rollouts promote rollout-canary
```

After running this command, the rollout will start performing an automated 20% increase gradually until it reaches 100%.

Enough with the theory, we have the fundamental knowledge to proceed to a real use case in the next section, which ties Argo CD and Argo Rollouts in a production CI/CD pipeline.

A real CI/CD pipeline

We discussed Argo Rollouts and how it works, and the deployment strategies, but how can we adapt all these in a real production environment? How can we integrate Argo Rollouts and automate the rollout without the need for manual approval? In this section, we will minimize failed deployments with Argo Rollouts and Argo CD and bootstrap our K8s cluster ready with Argo Rollouts.

Setting up Argo Rollouts

In *Chapter 5, Argo CD Bootstrap K8s Cluster*, we bootstrapped the EKS cluster we created with Terraform, and we will evolve this to include Argo Rollouts in the bootstrap. So, we will create another new Argo application for Argo Rollouts under the `ch05/terraform/k8s-bootstrap/base` directory. The following declarative manifest is the Argo application for bootstrapping the cluster with Argo Rollouts:

```
apiVersion: argoproj.io/v1alpha1
kind: Application
metadata:
  name: argo-cd
  finalizers:
    - resources-finalizer.argocd.argoproj.io
spec:
  project: default
  source:
    repoURL: https://github.com/PacktPublishing/ArgoCD-in-
Practice.git
    targetRevision: HEAD
    path: ch05/applications/argo-rollouts
  destination:
    namespace: argo-rollouts
    server: https://kubernetes.default.svc
  syncPolicy:
    automated:
      prune: true
```

```
      selfHeal: true
  syncOptions:
    - CreateNamespace=true
```

Finally, update `kustomization.yaml` to include the new application during the bootstrap:

```
namespace: argocd
bases:
  - https://raw.githubusercontent.com/argoproj/argo-cd/v2.1.7/
manifests/install.yaml
resources:
  - namespace.yaml
  - argocd.yaml
  - argo-rollouts.yaml
  - master-utilities.yaml
```

Now, it's time to apply the Terraform script:

```
$ terraform apply -auto-approve
```

So, now we have Argo Rollouts deployed in the cluster ready to start using it later in this section. Now, we will create an Argo Project so we can have the separation by the team and give the freedom to the team to manage their services by themselves in their own pipelines. After the script application, we need to check that the Argo application is healthy and synced, as we can see in *Figure 6.5*:

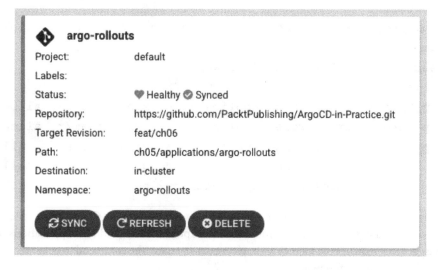

Figure 6.5 – Bootstrap cluster with Argo Rollouts and a demo app

In the next section, we will talk about team separation and implement a CI/CD pipeline with GitHub Actions.

Automated blue-green promotion with GitHub Actions

When we are in a big engineering team, we want freedom within our team to manage and deploy the Argo applications we need accordingly. That's why (as we discussed in *Chapter 2, Getting Started with Argo CD*, and *Chapter 4, Access Control*), we can use Argo Project and Project roles and tokens in the CI system we use. With this approach, we will fully utilize the multitenancy of Argo CD. We will assume that the project we will create for this case is for our team only:

```
apiVersion: argoproj.io/v1alpha1
kind: AppProject
metadata:
  name: team
spec:
  destinations:
  - namespace: team-*
    server: '*'
  sourceRepos:
  - https://github.com/PacktPublishing/ArgoCD-in-Practice.git
  roles:
  - name: team-admin
    policies:
    - p, proj:team:team-admin, applications, *, team/*, allow
  - name: ci-role
    description: Create and Sync apps
    policies:
    - p, proj:team:ci-role, applications, sync, team/*, allow
    - p, proj:team:ci-role, applications, get,  team/*, allow
    - p, proj:team:ci-role, applications, create, team/*, allow
    - p, proj:team:ci-role, applications, update, team/*, allow
    - p, proj:team:ci-role, applications, delete, team/*, allow
```

After creating this, we need to generate a Project token that we can use in the CI and, more specifically, use `ci-role`, which has limited permissions to create, update, delete, and sync applications under every namespace that matches the following simple `team-*` regex. The command to generate a token, for example, for a day is the following:

```
argocd proj role create-token team ci-role -e 1d
```

The output will be the following, and keep in mind that these tokens are not stored anywhere in Argo CD:

```
Create token succeeded for proj:team:ci-role.
ID: 3f9fe741-4706-4bf3-982b-655eec6fd02b
Issued At: 2021-12-12T00:05:41+02:00
Expires At: 2021-12-13T00:05:41+02:00
Token: <your-generated-token>
```

We will set this as a secret in GitHub Actions. Now is the time to define the CI/CD paths for a blue-green deployment so we can start implementing it later.

We are going to use the application we created in the previous section for blue-green deployment. For this example, we are going to use GitHub Actions as a pipeline, and we will define the following steps:

- Lint

- Build

- Deploy

We will set up two workflows with GitHub Actions. The first workflow will be defined in `ci.yaml` and will be responsible for creating artifacts (Docker images) for pull requests and the `main` branch. We are talking about the CI part of our pipeline that needs to run in every new change we raise in the repository. Here is the part that we define to run only on pull requests and the `main` branch:

```
name: main
on:
  push:
    branches: [ main ]
  pull_request:
    branches: [ '*' ]
```

Then, we have the `lint` step of the pipeline and the Jobs of the workflow:

```
jobs:
  lint:
    name: Lint
    runs-on: ubuntu-latest
    steps:
      ... more ...
      - name: Lint
        run: make lint
```

The last step is to build the Docker image and push it to Docker Hub, as shown in the following:

```
build:
  name: Build
  runs-on: ubuntu-latest
  needs: ["lint"]
  steps:
    ... more ...

    - name: Log in to Docker Hub
      uses: docker/login-action@v1
      with:
        username: ${{ secrets.DOCKER_USERNAME }}
        password: ${{ secrets.DOCKER_PASSWORD }}

    - name: Build and Push docker image
      run: |
        TAG=${BRANCH_NAME} make push-docker
      env: |
        BRANCH_NAME= ${{ steps.branch.outputs.current_branch
}}
```

With the next step, each team will be totally self-managed to deploy their services after code reviews by the relevant teams. Although in the case of a K8s cluster failure, it will be a bit harder to bootstrap a new K8s cluster or it will need some extra manual steps, so it's a good practice to always keep the right state of the Git repository used for the bootstrap. We will use the Argo CD CLI to create the Argo application under the team's project, and this will run only on the main branch. Here are the last steps:

```
- name: Download Argo CD CLI
  run: |
  make download-argo-cli

- name: Create Argo app
  run: |
  make create-argo-app
  env:
  PROJ: team
  APP: blue-green
```

```
        ROLE: ci-role
        JWT: ${{ secrets.ARGOCD_TOKEN }}
        ARGOCD_SERVER: ${{ secrets.ARGOCD_SERVER }}
        ARGOCD_PATH: "deployments/argo"
```

The make create-argo-app command uses the Argo CD CLI to create a new Argo app and sync it:

```
.PHONY: create-argo-app
create-argo-app:
@echo Deploying Argo App
argocd app create ${APP} \
--repo https://github.com/spirosoik/argocd-rollouts-cicd.git \
--path ${ARGOCD_PATH} \
--dest-namespace team-demo \
--dest-server https://kubernetes.default.svc \
--sync-option CreateNamespace=true \
--project ${PROJ} \
--auth-token ${JWT} \
--upsert
```

Lastly, if we had multiple clusters defined here, we could deploy the main branch to a dev cluster and tags to the production cluster in the CI/CD workflows. As long as the files under deployments/argo don't change, we will keep the old version served by both services as it's the first-time deployment. So, if we try to make a request, we will see the same version for both:

```
$ kubectl -n team-demo port-forward svc/rollout-bluegreen-
active 3000:80
$ curl localhost:3000 # In separate terminal
v1.0
```

Let's do the same for the preview service:

```
$ kubectl -n team-demo port-forward svc/rollout-bluegreen-
preview 3000:80
$ curl localhost:3000 # In separate terminal
```

Also, as we didn't promote the green version, the state of the Argo application will be suspended and we will get a message in the Argo CD events:

```
Rollout is paused (BlueGreenPause)
```

In *Figure 6.6*, you will see that the status of the Argo app is suspended:

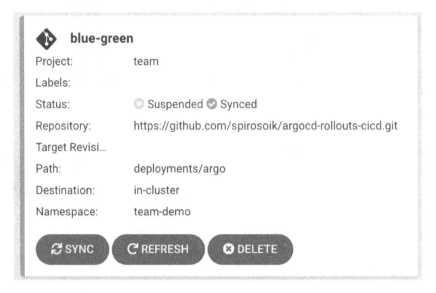

Figure 6.6 – Argo Rollouts suspended

The full example of the CI can be found at `https://github.com/PacktPublishing/ArgoCD-in-Practice` in the `ch06/automated-blue-green/.github/workflows/ci.yaml` folder.

The second workflow will be defined in `cd.yaml` and will be responsible for creating artifacts (Docker images) when we create new tags and then deploy using the blue-green strategy. Then, we will run a smoke test and if everything is alright, we will promote the green version to production deployment automatically.

The CD process for tags only is mostly the same, but there are two extra steps in the GitHub workflow:

- Deploy
- Smoke test

The deploy step is when there is a change of the tag that is served by the Argo Rollouts manifest, and it is replaced with the one we built. So, we change the tag and commit this change to the repository. Argo CD is responsible for observing the change and syncing everything to the proper state, the GitOps way:

```
- name: Download Argo CD CLI
  run: |
    make download-argo-cli


- name: Update Docker Tag
```

```
    run: |
      TAG="${GITHUB_REF##*/}" BUILD_NUMBER=${GITHUB_RUN_
NUMBER} make update-docker-tag

  - name: Deploy Argo App
    run: |
      make deploy-argo-app
    env:
      PROJ: team
      APP: blue-green
      ROLE: ci-role
      JWT: ${{ secrets.ARGOCD_TOKEN }}
      ARGOCD_SERVER: ${{ secrets.ARGOCD_SERVER }}
```

`deploy-argo-app` will sync and wait for the suspended state, as we discussed before:

```
@echo Syncing Argo App
argocd app sync ${APP} --auth-token ${JWT}

@echo Waiting Argo App to be healthy
argocd app wait ${APP} --auth-token ${JWT} --suspended
--timeout=120s
```

So far, we didn't run any blue-green switch, but we just deployed a new version, but we still have the old version active. Let's give it a try:

```
$ kubectl -n team-demo port-forward svc/rollout-bluegreen-
active 3000:80
$ curl localhost:3000 # In separate terminal
v1.0
```

Let's do the same for the `preview` service:

```
$ kubectl -n team-demo port-forward svc/rollout-bluegreen-
preview 3000:80
$ curl localhost:3000 #  In separate terminal
V2.0
```

Now, it's time to run the smoke tests and roll out the green version if everything is alright. The smoke test for the sake of the example will be just a URL to the services and validate that we get the v2.0 response back. In a production environment, most of the time, we have a set of integration tests that we can run in the green version and validate that the service is functional and ready.

Now, it's time to do the rollout and, to do this, we need the `kubectl` plugin for Argo Rollouts. But wait! For `kubectl`, we need `kubeconfig`, so we will expose, even with limited access, the credentials in the CI/CD system instead of respecting the pull approach of GitOps. But how can we avoid using `kubeconfig` in this case? For example, we need to run the following command to roll out:

```
kubectl argo rollouts promote app -n team-demo
```

The same question is relevant for smoke tests: what we will do if the services are only for internal access, and we are using a managed CI/CD system such as GitHub Actions? The answer again probably is that we need `kubeconfig` and to port-forward the services and run the tests, right?

Here comes the power of resource hooks and `sync-waves` in Argo CD, which will help us include everything as part of Argo CD and avoid having `kubeconfig` in the external CI/CD system. Let's see the implementation in the next section.

Automated rolling out with sync phases

In *Chapter 2, Getting Started with Argo CD*, we discussed resource hooks and how we can create some workflows based on the sync phases. As a reminder, *Figure 6.7* will remind you of how sync phases work:

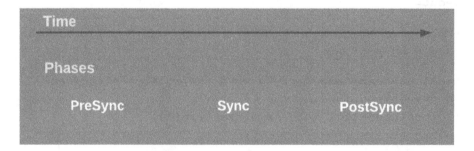

Figure 6.7 – Argo CD sync phases

In our case, we will run the integration tests as a separate Argo CD app and will use `Sync` and `PostSync` phases. In the `Sync` phase, we will run the integration tests and if the sync has been completed with a healthy state, then will proceed to the `PostSync` phase, where we will promote the new version. With this practice, we minimize the failed deployments with Argo Rollouts and smoke tests. If the tests fail, `PostSync` will never run, and by utilizing Argo CD resource hooks, we will keep the containers to debug the issue. What we described is the superpower of Argo CD combined with other Argo Projects such as Argo Rollouts.

Let's see the manifest we will use to run the integration tests with just a simple smoke test:

```
apiVersion: batch/v1
kind: Job
metadata:
  generateName: integration-tests
  namespace: team-demo
  annotations:
    argocd.argoproj.io/hook: Sync
    argocd.argoproj.io/hook-delete-policy: HookSucceeded
spec:
  template:
    spec:
      containers:
      - name: run-tests
        image: curlimages/curl
        command: ["/bin/sh", "-c"]
        args:
          - if [ $(curl -s -o /dev/null -w '%{http_
code}' rollout-bluegreen-preview/version) != "200" ]; then
exit 22; fi;
            if [[ "$(curl -s rollout-bluegreen-preview/
version)" != "APP_VERSION" ]]; then exit 22; fi;
            echo "Tests completed successfully"
      restartPolicy: Never
  backoffLimit: 2
```

The main part here is the two annotations we have in the K8s job:

```
argocd.argoproj.io/hook: Sync
argocd.argoproj.io/hook-delete-policy: HookSucceeded
```

These two indicate that the job will run in the Sync phase and if it successfully completed, the job will be automatically deleted by the HookSucceeded resource hook. The last part is to roll out the app in the PostSync phase. Here is the job for rolling out the new version:

```
apiVersion: batch/v1
kind: Job
metadata:
```

```
    generateName: rollout-promote
    namespace: team-demo
    annotations:
      argocd.argoproj.io/hook: PostSync
      argocd.argoproj.io/hook-delete-policy: HookSucceeded
  spec:
    template:
      spec:
        containers:
        - name: promote-green
          image:  quay.io/argoproj/kubectl-argo-rollouts:v1.1.1
          command: ["/bin/sh", "-c"]
          args:
            - kubectl-argo-rollouts promote app -n team-demo;
        restartPolicy: Never
    backoffLimit: 2
```

The main part here is the two annotations we have in the K8s job:

```
argocd.argoproj.io/hook: PostSync
argocd.argoproj.io/hook-delete-policy: HookSucceeded
```

So, the GitHub Action CD is pretty simple; check *Figure 6.8* for what it looks like:

Figure 6.8 – GitHub actions pipeline

You can see the full list of steps of the **Deploy** job in *Figure 6.9*:

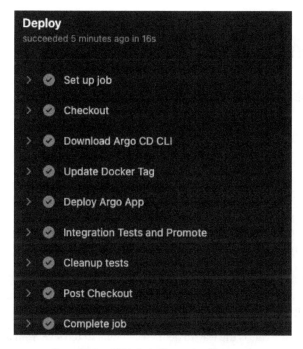

Figure 6.9 – GitHub Actions pipeline steps

That was an awesome journey to build a close-to-reality CI/CD pipeline utilizing Argo CD and Argo Rollouts and minimizing the failed deployments with Argo CD Sync phases. Now it's time to jump to the next section, where we will see how we can keep the secrets safe in the Git repository when we apply GitOps practices.

Keeping secrets safe

As we are talking about GitOps and declarative configuration (K8s manifests, Helm, and so on) in a Git repository, the first problem we need to address is how we can store the secrets safely. Let's see how we can achieve this in GitOps.

Storing secrets safely

The most secure way to store them is to keep them in a secret management tool such as Vault, AWS Secrets Manager, Azure Key Vault, or Google's Secret Manager. But how can you do this integration with Kubernetes Secrets and a declarative manifest and utilize GitOps practices?

There is a tool called *External Secrets Operator*. As the K8s operator is designed for automation, External Secrets Operator more specifically will synchronize secrets from external APIs such as AWS Secret Manager, Vault, and a couple of others into Kubernetes Secret resources.

The whole idea is that there are a few new K8s custom resources that will define where the secret is and how to complete the synchronization. Let's see the resource data model so that we can understand the mechanics of it a bit more in *Figure 6.10*:

Figure 6.10 – Resource model

The components of the resource model are the following:

- `SecretStore`: This is the authentication part of the external API so that we can retrieve the actual secret. It will check for new secret resources under the same namespace where it's created. Finally, it can be referenced only in the same namespace.

- `ExternalSecret`: This is the way to define what data to retrieve from an external API, and it interacts with `SecretStore`.

- `ClusterSecretStore`: This is a global secret and can be referenced by any namespace in the cluster.

An example of `SecretStore` is the following manifest:

```
apiVersion: external-secrets.io/v1alpha1
kind: SecretStore
metadata:
  name: secretstore-sre
spec:
  controller: dev
  provider:
    aws:
      service: SecretsManager
      role: arn:aws:iam::123456789012:role/sre-team
      region: us-east-1
      auth:
        secretRef:
          accessKeyIDSecretRef:
```

```
      name: awssm-secret
      key: access-key
    secretAccessKeySecretRef:
      name: awssm-secret
      key: secret-access-key
```

The relevant `ExternalSecret` resource that will use the `SecretStore` example is the following:

```
apiVersion: external-secrets.io/v1alpha1
kind: ExternalSecret
metadata:
  name: db-password
spec:
  refreshInterval: 1h
  secretStoreRef:
    name: secretstore-sre
    kind: SecretStore
  target:
    name: secret-sre
    creationPolicy: Owner
  data:
  - secretKey: dbPassword
    remoteRef:
      key: devops-rds-credentials
property: db.password
```

The example we have is separated by teams, which is a nice logical separation of your secrets so you can group the secrets of the team. This, of course, is not ideal for every team because you may want to keep some secrets safe only for specific people.

The idea behind the external secrets is similar to what we discussed with Argo about reconciliation and keeping the state always synced but, in this case, the External Secrets controller will update the secrets every time they change in a particular secret manager in order to keep the same state.

Next, we will discuss how we can use External Secrets Operator along with Argo CD, and the benefits of using it.

Argo CD and External Secrets Operator

Argo CD relies on Git state, and we cannot save plain passwords or sometimes even encrypted passwords in our repositories. External Secrets Operator comes in to solve this problem and it has several advantages, such as the following:

- Minimizes security risks

- Fully automated

- Use the same tool across every environment

All these are nice but how can I solve the problem of creating the secrets that I need first and then the application that will use them? Of course, again, here sync-waves and sync phases come in as we discussed earlier in this chapter to give the solution and orchestrate the order of the deployments so we can do this.

In order to bootstrap a cluster and have the operator before every other deployment because we may need it to get a secret, we need to use sync-waves so it can be installed first.

Let's assume that we want to deploy, for example, Grafana, an nginx to access it, and finally, we need to get a password for the admin; we will follow this in *Figure 6.11*:

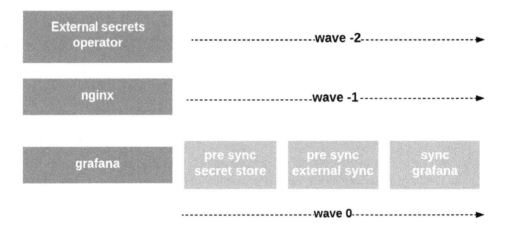

Figure 6.11 – Argo CD, External Secrets with sync-waves and phases

We then leverage the order of the waves and the PostSync phase to first create SecretStore and ExternalSecret in the order the Kubernetes Secret will be created to be ready for use by Grafana.

Time to see another example of a microservice CI/CD pipeline with multiple K8s and how Argo CD comes in to solve this problem.

Microservices CI/CD

In this section, we will look at how the new Argo `ApplicationSet` controller (which we described in *Chapter 5, Argo CD Bootstrap K8s Cluster*) evolves the pattern to be more flexible and support more use cases such as *monorepos* with microservices, which we will discuss in the next section.

Monorepo microservices

Many companies out there host everything under one monorepo instead of creating multiple repositories for each microservice and sometimes infrastructure scripts in addition to microservices. Argo CD with `sync-waves` and `ApplicationSet` makes the orchestration and the deployment of multiple services easy.

First, let's install the `ApplicationSet` controller in the cluster by bootstrapping it. In the *Deployment strategies* section earlier in this chapter, we created multiple Argo applications for the following:

- Master Utilities, which is an App of Apps pattern

- Argo Rollouts, which is an Argo application

- Blue-green application

Imagine that we can have a couple more here that will grow significantly. Let's change this with the `ApplicationSet` CRD. First, we will simplify the `kustomize` manifest to this:

```
apiVersion: kustomize.config.k8s.io/v1beta1
kind: Kustomization
namespace: argocd
bases:
  - https://raw.githubusercontent.com/argoproj/argo-cd/v2.1.7/
manifests/install.yaml
resources:
  - namespace.yaml
  - argo-applicationset.yaml
  - bootstrap-applicationset.yaml
```

The difference is that we removed all the other resources and added two new ones: `argo-applicationset.yaml` and `bootstrap-applicationset.yaml`. The first part is related to installing the Argo `ApplicationSet` controller in the cluster. The second part is the transformation from many apps to `ApplicationSet` and using the generators to exclude some directories and include a specific cluster. See the following `ApplicationSet` manifest:

```yaml
apiVersion: argoproj.io/v1alpha1
kind: ApplicationSet
metadata:
  name: bootstrap
spec:
  generators:
    - matrix:
        generators:
          - git:
              repoURL: https://github.com/PacktPublishing/ArgoCD-in-Practice.git
              revision: HEAD
              directories:
                - path: ch05/applications/*
                - path: ch05/applications/istio-control-plane
                  exclude: true
                - path: ch05/applications/argocd-ui
                  exclude: true
          - list:
              elements:
                - cluster: engineering-dev
                  url: https://kubernetes.default.svc
  template:
    metadata:
      name: '{{path.basename}}'
    spec:
      project: default
      source:
        repoURL: https://github.com/PacktPublishing/ArgoCD-in-Practice.git
        targetRevision: HEAD
        path: '{{path}}'
```

```
     destination:
        server: '{{url}}'
        namespace: '{{path.basename}}'
```

We combined two different generators with a matrix generator, and we are leveraging the ability to create applications while iterating the directories under `ch05/applications/*`. This is awesome as now we don't need to create an Argo application for each one, just the `ApplicationSet` controller, which is responsible for doing it automatically for us.

Let's assume that we have a microservices *monorepo* where we have multiple microservices and, in parallel, multiple clusters to deploy them with the following structure, and there are a few dependencies to each other microservices (like they need to start in a specific order):

```
|- service1
     |
     -- helm
          |-- values-dev.yaml
          |-- values-staging.yaml
     -- application
|- service2
     |
     -- helm
          |-- values-dev.yaml
          |-- values-staging.yam
     -- application
|- service3
     |
     -- helm
          |-- values-dev.yaml
          |-- values-staging.yam
     -- application
```

The `ApplicationSet` controller CRD should be defined as in the following:

```
apiVersion: argoproj.io/v1alpha1
kind: ApplicationSet
metadata:
  name: bootstrap
spec:
```

```
generators:
  - matrix:
      generators:
        - git:
            repoURL: https://github.com/PacktPublishing/
ArgoCD-in-Practice.git
            revision: HEAD
            directories:
              - path: '*'
        - list:
            elements:
            - cluster: dev
              url: https://kubernetes.default.svc
            - cluster: staging
              url: https://9.8.7.6
template:
  metadata:
    name: '{{path.basename}}'
  spec:
    project: default
    source:
      repoURL: https://github.com/PacktPublishing/ArgoCD-in-
Practice.git
      targetRevision: HEAD
      path: '{{path}}/helm'
      helm:
        valueFiles:
          - values-{{cluster}}.yaml
    destination:
      server: '{{url}}'
      namespace: '{{path.basename}}'
```

The last thing is to solve the dependencies between each other microservice where, again, we are relying on sync-waves of Argo CD so we can respect the microservices' interdependencies, as *Figure 6.12* represents:

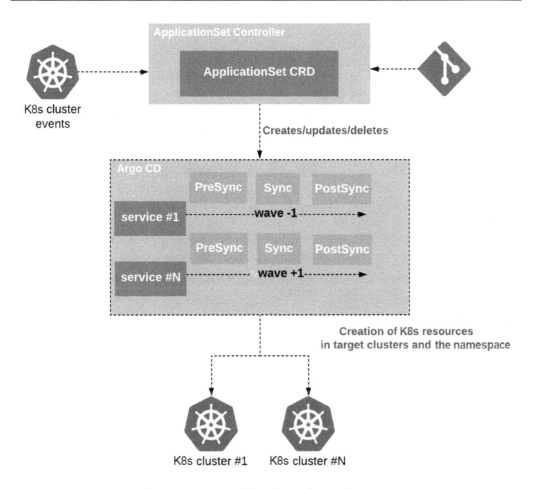

Figure 6.12 – Microservices CI/CD with ApplicationSet in a monorepo

We see again that we utilize `sync-waves` to order the deployment of the services and for each of them, we use the `Sync` phases to end up in a healthy state and continue to the deployment of the next service.

We just checked how we can use `ApplicationSet` and the great power it gives us to avoid too many apps and the easy way to deploy with the DRY approach in multiple clusters. Now it's time to summarize what we have learned and wrap up the most practical chapter so far with CI/CD.

Summary

In this chapter, we learned about the deployment strategies we can follow with Argo CD combined with Argo Rollouts and how we can minimize deployment failures with automation. We created a real CI/CD pipeline in GitHub Actions, and we implemented a full blue-green deployment in Argo CD with GitOps practices. We did a logical group separation in Argo CD with Argo Projects, and we gave limited to access with the Project tokens in a CI/CD system.

We leveraged `Sync` phases and resource hooks in order to run integration tests and fail the pipeline if those tests failed and most importantly, keep the resource to debug the reasons for the failure in the `PreSync` phase. The hook policy of Argo CD gave us the power to delete the redundant completed-with-success integration tests so we could keep our cluster clean. In the `PostSync` phase, when all phases ran successfully, we were able to roll out the latest version of our application without any failures. Also, we used `ApplicationSet` to use one single K8s manifest and deploy with the DRY approach to multiple clusters and multiple applications and support easier monorepos for microservices.

We hoped to achieve having a CI/CD pipeline close to production and to inspire you to approach your challenges safely.

Further reading

- Argo Rollouts: `https://argoproj.github.io/argo-rollouts/`
- Minimize failed deployments with Argo Rollouts and smoke tests: `https://codefresh.io/continuous-deployment/minimize-failed-deployments-argo-rollouts-smoke-tests/`
- Canary CI/CD: `https://github.com/codefresh-contrib/argo-rollout-canary-sample-app`
- Traffic management: `https://argoproj.github.io/argo-rollouts/features/traffic-management/`
- ApplicationSet: `https://argocd-applicationset.readthedocs.io/en/stable/`

7
Troubleshooting Argo CD

In this chapter, we will see how to address some of the issues you are most likely to run into during your day-to-day work with Argo CD. We will go through the installation procedure and see what can happen if we try to have more than one installation in the same cluster. This is a valid scenario for organizations that have many clusters deployed in **development** (**dev**) and **production** (**prod**) and need to split the load to different Argo CD instances. Then, we will discover how to use a specific version of Helm, different than the one embedded in the Docker image. Sometimes, this is needed because a newer version of a templating engine can bring some breaking changes and you would need more time to prepare for the upgrade, but you still want to have the performance improvements of the new Argo CD version. Argo CD has become much more stable in recent years—it is a mature application used by many organizations in production, but there are still cases when a good old restart is needed, so we will see how to perform this correctly. We will finish the chapter with some explanations around performance, and go through and explain in detail those configuration parameters and environment variables that can make a difference to overall performance.

In this chapter, we're going to cover the following main topics:

- Initial setup
- Addressing day-to-day issues
- Improving performance

Technical requirements

For this chapter, we will start with a new installation of Argo CD (actually, more than one), but it will not be in the `argocd` namespace, so it will be a fresh and different start.

We are going to write quite a lot of **YAML Ain't Markup Language** (**YAML**), so a code editor will be needed, and as usual, I'll go with **Visual Studio Code** (**VS Code**) (`https://code.visualstudio.com`). All the changes we will make can be found at `https://github.com/PacktPublishing/ArgoCD-in-Practice` in the ch07 folder.

Initial setup

Almost all of the tutorials, blogs, and articles out there discuss installing Argo CD in the `argocd` namespace, and it does make sense as it is a specific namespace—it will not collide with other applications. But there are cases when that might not be possible—I am talking about when multiple instances of Argo CD are installed in the same Kubernetes cluster. This is a valid scenario when we want to split clusters that an Argo CD instance is managing, such as one that takes care of prod clusters and another the non-prod ones. Another scenario could be when they take care of the same clusters but only specific namespaces from it, so a model of an Argo CD instance per team. We will follow the scenario with prod and non-prod installations.

When we have two or more Argo CD installations in the same cluster, there are a couple of things we need to pay attention to. First is the fact that when we install it in the cluster, there are some **Custom Resource Definitions** (**CRDs**) (`https://kubernetes.io/docs/concepts/extend-kubernetes/api-extension/custom-resources/`) being applied to allow us to create Applications and AppProjects resources. These definitions should be created only once, so only one Argo CD instance should apply them. So, we need to decide which of the two instances—prod or non-prod—will take care of the cluster where we actually installed them (as the CRDs will be installed on that cluster).

Usually, the job of **site reliability engineers** (**SREs**) is to create a self-service platform for developers in order to deploy their work to production, so most likely we will need to run more applications, not just Argo CD. We might need **Prometheus Operator** (`https://prometheus-operator.dev`), maybe **ChartMuseum** (`https://chartmuseum.com`) for storing Helm charts, or something like Harbor (`https://goharbor.io`) if we want to host our own container registry, or even host our own **continuous integration/continuous deployment** (**CI/CD**) system such as Tekton (`https://tekton.dev`). We will run all these applications in the same cluster, and we usually name it the Platform cluster. In order to deploy all these, we can have a third Argo CD instance that will take care of everything we want to set up in this Platform cluster, including applying the Argo CD CRDs. You can see what a setup involving three Argo CD instances in the same cluster would look like in the following diagram—a prod instance that takes care of all our production clusters, a non-prod instance that takes care of our dev, test, and staging clusters (this will be our test instance where we will verify the upgrades and any other configuration changes), and—finally—a Platform instance that takes care of the Platform cluster (including the installation of prod and non-prod instances):

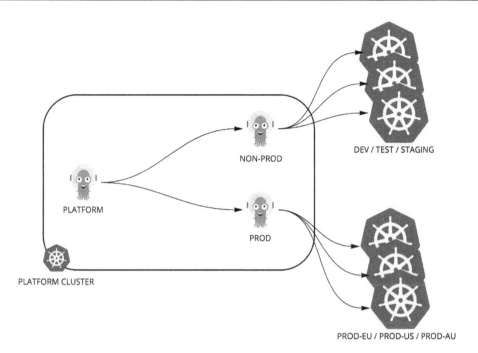

DEV / TEST / STAGING

NON-PROD

PLATFORM

PROD

PLATFORM CLUSTER

PROD-EU / PROD-US / PROD-AU

Figure 7.1 – Multiple Argo CD instances on the same cluster

First, let's deploy the Argo CD Platform instance, which is a complete installation, including the CRDs, and we will use Kustomize to deploy it. We can install it in an imperative way by first running `kubectl apply` to create a namespace and then using `kustomize build` to apply the remote resources from the Argo CD GitHub repository. But it is better to use a declarative approach because we can also enable the self-management part, which means the resources for all three installations should be stored in a Git repository (or more than one) to be able to update them via **pull requests** (**PRs**).

There are two options here—have all the installation manifests in the same repository on different branches named platform, prod, non-prod, or store all of them in separate folders in the same branch. When working with different branches, you will always need to pay attention to which one you open the PRs to as they go to the default branch, so mistakes could happen. That's why it's maybe easier to go with one Git repository with a folder for each installation. In our case, for simplicity, we will have one folder for each under the `ch7/initial-setup` folder in the official repository `https://github.com/PacktPublishing/ArgoCD-in-Practice` (but normally, they should be under `root`). We are starting with installing the initial Platform instance, which will be under `ch7/initial-setup/argocd-platform`.

We will create a `resources` folder and, inside it, a `namespace.yaml` file with the following content:

```
apiVersion: v1
kind: Namespace
```

```
metadata:
  name: argocd-platform
```

Outside the `resources` folder, we will create a `kustomization.yaml` file, like so:

```
apiVersion: kustomize.config.k8s.io/v1beta1
kind: Kustomization
namespace: argocd-platform
bases:
  - github.com/argoproj/argo-cd/manifests/ha/cluster-
install?ref=v2.1.6
resources:
  - resources/namespace.yam
```

From the folder where the `kustomization.yaml` file resides (when you have one repository per instance that should be in the root of the repository), run the following command to make the installation:

```
kustomize build . | kubectl apply -f -
```

After everything has been applied, we can run this command to see the running Pods in the newly created namespace, in order to check that our installation was done:

```
kubectl get pods -n argocd-platform
```

The result should be a list of Pods similar to this:

NAME	READY	STATUS	RE-STARTS	AGE
argocd-application-controller-0	1/1	Running	0	4m
argocd-dex-server-6bf7f6876c-xghc6	1/1	Running	0	4m
argocd-redis-5b6967fdfc-m5qff	1/1	Running	0	4m
argocd-repo-server-685cb7dbdd-qk6xp	1/1	Running	0	4m
argocd-server-bfb77d489-tbqww	1/1	Running	0	4m

Now that we have the Platform installation, we can start deploying the prod and non-prod ones (prod will be under `ch7/initial-setup/argocd-prod`, while non-prod will be under `ch7/`

`initial-setup/argocd-nonprod`). For this part of the demo, we will provide instructions only for the non-prod installation, as the prod one should follow the same steps. We will continue with the declarative way and the `namespace.yaml` file under the `resources` folder and the `kustomization.yaml` file outside, but this time we will apply them with an `Application` **custom resource** (**CR**) because we already have an instance that can handle them.

Here is the `namespace.yaml` file for non-prod:

```
apiVersion: v1
kind: Namespace
metadata:
  name: argocd-nonprod
```

We will create a `kustomization.yaml` file where we will not use the `cluster-install` manifests but instead the `namespace-install` ones. We will see the differences between such installations right after we finish the installation. Here's the code we need to execute:

```
apiVersion: kustomize.config.k8s.io/v1beta1
kind: Kustomization
namespace: argocd-nonprod
bases:
  - github.com/argoproj/argo-cd/manifests/ha/namespace-
install?ref=v2.1.6
resources:
  - resources/namespace.yaml
```

Then, we create an `Application` file that will point to the repository and folder where we will have the `kustomization.yaml` file. I named the file `argocd-nonprod-app.yaml`, and it should look like this:

```
apiVersion: argoproj.io/v1alpha1
kind: Application
metadata:
  name: argocd-nonprod
spec:
  destination:
    namespace: argocd-nonprod
    server: https://kubernetes.default.svc
  project: default
  source:
```

```
    path: ch7/initial-setup/argocd-nonprod
    repoURL: https://github.com/PacktPublishing/ArgoCD-in-
Practice.git
    targetRevision: main
  syncPolicy:
    automated:
      prune: false
```

Once we have all those files created, we can apply the `Application` file with `kubectl`, using the following command:

```
kubectl apply -f argocd-nonprod-app.yaml -n argocd-platform
```

Unfortunately, the result will not be the expected one, and we will get an error similar to this one:

```
failed to sync cluster https://10.96.0.1:443: failed to
load initial state of resource CertificateSigningRequest.
certificates.k8s.io: certificatesigningrequests.certificates.
k8s.io is forbidden: User "system:serviceaccount:argocd-
platform:argocd-application-controller" cannot list resource
"certificatesigningrequests" in API group "certificates.k8s.io"
at the cluster scope
```

This is what the details page of the `argocd-nonprod` `Application` file will look like. See the **2 Errors** red text—if you click on it, you will get a similar error message to the previous one:

Figure 7.2 – Errors while applying argocd-nonprod Application file

The problem comes from the fact that in the Kustomize cluster installation, the namespace is hardcoded to argocd for the ClusterRoleBinding resources for both application controller and server components. Because of this, the Argo CD Platform installation does not have access to the resources it needs to watch and create in the cluster, and so it gives the previous error. So, if we can update those ClusterRoleBnding manifests, everything should start working.

> **Note – Cluster versus Namespace installation with Helm**
>
> When using Helm manifests for the installation, two parameters control the cluster **role-based access control** (RBAC) installation: .Values.server.clusterAdminAccess.enabled and .Values.controller.clusterAdminAccess.enabled. They are enabled by default, so you need to set them to false when you want to perform a namespace installation, and we don't have the same issue when performing a cluster install in a different namespace than argocd because the namespace is templated to namespace: {{ .Release. Namespace }}.

The change we are doing is in for the Argo CD Platform installation, and we are updating two ClusterRoleBinding resources in order to use the ServiceAccount instances from the argocd-platform namespace. We will create a patches folder under argocd-platform, and in this new folder, we will create a file called argocd-application-controller-clusterrolebinding.yaml with this content:

```
apiVersion: rbac.authorization.k8s.io/v1
kind: ClusterRoleBinding
metadata:
 name: argocd-application-controller
subjects:
- kind: ServiceAccount
 name: argocd-application-controller
 namespace: argocd-platform
```

We will need to do the same for the argocd-server ServiceAccount instance. We need another file called argocd-server-clusterrolebinding.yaml in the patches folder, having this content:

```
apiVersion: rbac.authorization.k8s.io/v1
kind: ClusterRoleBinding
metadata:
 name: argocd-server
subjects:
- kind: ServiceAccount
```

```
  name: argocd-server
  namespace: argocd-platform
```

In order for the files to be used by our installation to perform the namespace change, we need entries for them in the kustomization.yaml file. It should look like this:

```
apiVersion: kustomize.config.k8s.io/v1beta1
kind: Kustomization
namespace: argocd-platform
bases:
 - github.com/argoproj/argo-cd/manifests/cluster-
install?ref=v2.1.6
resources:
 - resources/namespace.yaml
patchesStrategicMerge:
  - patches/argocd-application-controller-clusterrolebinding.
yaml
  - patches/argocd-server-clusterrolebinding.yaml
```

Now, if we apply the Kustomize manifests again, as follows, then we should get a correct installation under the argocd-platform namespace:

```
kustomize build .  | kubectl apply -f -
```

Now, we have the ClusterRoleBinding resources linked to the ServiceAccount instance from the correct namespace; this allows Argo CD to list those resource types, so we should not get the error anymore when we install the non-prod or prod instances.

After the initial setup is done and we have all our instances of Argo CD up and running, we are ready to connect to clusters and deploy applications. But sometimes, things will not go according to plan, and we might need to perform some basic restarts, or we might need to use specific versions of our templating engine. Next, we are going to cover these two scenarios.

Addressing day-to-day issues

In this section, we will discuss a few issues you would encounter on what we will call *Day 2*. If we mark *Day 0* as the moment when we prepare the GitOps and Argo CD adoption and start working on **proofs of concept** (**POCs**) and *Day 1* when we actually start implementing and deploying our GitOps solution, then *Day 2* would be when we are in production, and we need to take care of the whole live setup. We went through important topics such as **high availability** (**HA**), **disaster recovery** (**DR**), and observability in *Chapter 3*, *Operating Argo CD*, and now we will see some issues that don't fall into any of these categories.

Restarting components

Back in 2018 when we started with Argo CD, we had cases when we deployed a new version of one of our microservices and the synchronization just got stuck. The **user interface** (**UI**) showed that the reconciliation loop was ongoing, but nothing was happening, there was no progress, and no error was displayed. That's when a good old restart of the application controller would fix the issue. In time, with every new version released, Argo CD became more stable and reliable, and we had fewer instances of situations where a restart was needed.

Because every business demand and every environment are different, and an application built with Python doesn't start or stop the same way as one built with Java, a restart might still save the day from time to time. Let's see how that should be properly executed without getting Pods or—even worse—Deployments or StatefulSets deleted.

If you have access to the cluster, there is a simple command you can execute in order to trigger a restart, and that is `kubectl rollout restart`. If you encounter issues with synchronizations, then you can try to restart the application controller. This is usually deployed in Kubernetes as a StatefulSet, and the command would look like this (assuming the installation was done in the `argocd` namespace and the StatefulSet name is `argocd-application-controller`, which is the default value):

```
kubectl rollout restart sts argocd-application-controller -n
argocd
```

If you are using Helm to install Argo CD, the Helm chart uses a Deployment for the application controller by default (the `controller.enableStatefulSet` parameter is `false`). In this case, the command would look like this:

```
kubectl rollout restart deploy argocd-application-controller -n
argocd
```

The repository servers we would restart when encountering issues with generating the manifests and the command would look like this:

```
kubectl rollout restart deploy argocd-repo-server -n argocd
```

For the server, the restart could be useful when the **command-line interface** (**CLI**) or the UI stops responding because the server has some issues. In such cases, use the following command:

```
kubectl rollout restart deploy argocd-server -n argocd
```

We could restart Dex if we have problems with the single sign-on process, like so:

```
kubectl rollout restart deploy argocd-dex-server -n argocd
```

A Redis restart is not recommended as that would throw away the cache, and there would be a performance penalty because a lot of manifests need recalculation.

In a normal situation, you should try to understand why the process was stuck, why Argo CD is not responding, and address the underlying issue. But sometimes, there is no time to deal with that and we need a fast solution to unblock the situation, which then allows us to gain some time to start debugging without someone being blocked.

Using a specific version of Helm

Argo CD has the main templating tools inside its own container image. So, each new version of Argo CD comes bundled with a specific version of Helm or Kustomize. For example, version 2.0 has Helm 3.5.1 and Kustomize 3.9.4 (a complete list of tools and their version can be found at `https://github.com/argoproj/argo-cd/blob/v2.0.0/hack/tool-versions.sh`), while version 2.1 came with Helm 3.6 and Kustomize 4.2 (a complete list can be seen at `https://github.com/argoproj/argo-cd/blob/v2.1.0/hack/tool-versions.sh`). We have situations when our manifests depend on a specific version of the templating tool we are using. We might need more time to upgrade from Kustomize 3 to 4, but we do want to start using Argo CD 2.1 because of all the bug fixes and improvements it brings. Or, it can be the other way around, where we might need a newer version of a templating tool that has a fix for a bug that created problems for us recently.

One way to use specific versions of any tool is to create your own Argo CD container image. This way, you have control of everything, you decide which version is suited for your YAML templates, and you can even remove some of the bundled tools if you know you don't need them, thus resulting in a smaller container image. The problem with this approach is that it requires quite a bit of work: you need to build the image, push it to a container registry, and then do this for any new patch or minor versions—and there can be a few. Upgrading Argo CD is a pretty simple task, especially when using a self-managed instance, but now with your own container image, it can become a burden.

There is an easier way to use a specific version of Helm or any other templating tool. We can use an `init` container in our repository server deployment that will download the tool binary and then replace it in the container image. It might sound complicated, but it is pretty simple to set up, and using this method, we can upgrade Argo CD following the same process as always, because the `init` container part is separated from the main container. Next, we will see how we can do this with a Helm installation of Argo CD 2.1, which comes with Helm 3.6, but instead, we would like to have Helm 3.5.1 deployed. For how this can be accomplished for a Kustomize installation, we have the details in the official documentation at `https://argo-cd.readthedocs.io/en/stable/operator-manual/custom_tools`.

We can go through the process and see what it would look like when the installation is done via Helm and what would be the overwrites needed for the `values.yaml` file.

First, we need to define a new volume, which will be called `custom-tools` and will be of type `emptyDir`. This type of volume uses the default storage type used by the node, but we should not worry about the size as the file is around 40 **megabytes (MB)**. If the container is restarted, we will not have any problems as the volume contents are not lost—this only happens when the Pod is removed from the node.

Once we have the volume, we will mount it to both the `init` container and to the main container where all the Argo CD files are located. When we mount it to the main container, it is important that we specify the `subPath: helm` entry because this assures that we remove the previous file (the version of Helm we want to replace) and that the `mountPath` is set to `/usr/local/bin/helm`, which is the new file that we will retrieve in the `init` container.

In the `init` container, which is an `alpine:3.15` image, we are mounting the same volume and downloading the desired version of Helm into it using `wget`.

Create a new file called `values-overwrite.yaml` and put the following content into it (the file can also be found in the official repository under the `ch7/day-to-day/custom-helm/values-overwrite.yaml` path):

```
repoServer:
  volumes:
    - name: custom-tools
      emptyDir: {}
  volumeMounts:
    - mountPath: /usr/local/bin/helm
      name: custom-tools
      subPath: helm
  initContainers:
    - name: download-tools
      image: alpine:3.15
      command: [sh, -c]
      args:
        - wget -qO- https://get.helm.sh/helm-v3.5.1-linux-
amd64.tar.gz | tar -xvzf - && mv linux-amd64/helm /custom-
tools/
      volumeMounts:
        - mountPath: /custom-tools
          name: custom-tools
```

If you want to test the update, you can easily install Argo CD via Helm with this command:

```
helm install argocd argo/argo-cd -n argocd -f values-overwrite.
yaml
```

We did the example for Helm, but you can see that it would not be hard to modify it to use a specific version of Kustomize, the only changes being under the `init` container download part and mounting the volume in the correct path in the main container.

You can also see that the main container is separated from the `init` one, the only common point being the volume where we are downloading the new binary. This is why it is possible to upgrade Argo CD to the next patch version and still keep the benefits of having a fixed version of your template engine.

Improving performance

Throughout the book, we have touched on the topic of performance a few times by explaining some of the settings or environment variables used to get better results from an Argo CD installation. For example, in *Chapter 3*, *Operating Argo CD*, we explained the `ARGOCD_EXEC_TIMEOUT` environment variable for repository servers, which you can use to increase the time allocated for Helm or Kustomize to generate manifests so that you will not receive timeouts. In the same chapter, we discussed setting resources and increasing the number of replicas for all components and how this can be achieved. We are not going through those things again; instead, we are going to take a close look at some parameters we can use in order to control the application controller and repository server performance. These are flags that need to be set on the command that starts the containers.

Application controller

The application controller is the component where the work is being initiated. Whenever you click the **Sync** button from the UI, use the `argocd app sync` CLI command, or make an **application programming interface** (**API**) call directly to the server to start a synchronization or when poll timeout happens (by default, every 3 minutes), in the end, it is the application controller that performs the synchronization procedure, and we have some parameters that we can pass to its container at the start in order to influence this work. We will discuss them one by one, as follows:

- `--status-processors`—This is the parameter that specifies how many applications can be reconciled at the same time, concurrently. So, it makes sense that if you have an instance handling many applications, you would have a bigger value for it. The official documentation mentions the value of 50 for 1,000 applications, the default value being 20.

- `--operation-processors`—This has a default value of 10 and allows us to specify how many concurrent sync operations we can have. For 1,000 applications, it is recommended to go with 25.

- `--repo-server-timeout-seconds`—The application controller passes the job of generating manifests to the repository server containers. This takes time, as in many cases it needs to fetch artifacts from the internet (such as Helm downloading a chart from a registry or Kustomize downloading a remote base), and this is how much the application controller waits for the repository server to generate manifests before failing with a `Context deadline exceeded` exception. So, if this happens, increase the value of the parameter, the default being 60 seconds.

- `--kubectl-parallelism-limit`—After manifest generation is complete, we need to apply the manifests to the destination clusters. When we explained the GitOps concept in *Chapter 1, GitOps and Kubernetes*, we saw this is done declaratively with a `kubectl apply` command. This parameter is used to specify how many concurrent `kubectl` executions we can have, and the default is 20. You can increase it if you want your applications to be synchronized faster, but a big value might start triggering **out-of-memory** (**OOM**) kills for the container, so pay attention to the metrics.

For setting different values for these parameters on a Helm installation, you have them parameterized under `controller.args` or use the `controller.extraArgs` array, so the `values.yaml` file would look like this (I only show the section that corresponds to these parameters, while the complete override can be found in the official repository at `https://github.com/PacktPublishing/ArgoCD-in-Practice`, in the `ch7/performance-improvement/helm-installation/override-values.yaml` file):

```
controller:
  args:
    statusProcessors: "50"
    operationProcessors: "25"
    repoServerTimeoutSeconds: "60"
  extraArgs:
    - --kubectl-parallelism-limit=30
```

> **Note – Application Controller Additional Arguments**
>
> There are some additional flags present in the Helm values file: `appResyncPeriod` was deprecated in favor of the `timeout.reconciliation` setting in the `argocd-cm` ConfigMap, and `selfHealTimeout`, which is related to changes in live-cluster-triggering synchronizations, so it doesn't necessarily affect performance.

Now that we have discovered the parameters that can influence the performance of application controllers, we can check the ones related to repository servers.

Repository server

The repository server is responsible for generating manifests that can simply be applied to the destination cluster with a `kubectl apply` command. So, it will take Helm, Kustomize, or Jsonnet templates and generate final YAML manifests from them. Generated manifests are also cached for a 24-hour period on the basis that if the commit **secure hashing algorithm** (**SHA**) of the Git repository did not change, there is no need to regenerate manifests, and the cached version can be used. Here are some flags and environment variables that we can pass to the repository server container to influence manifest generation:

- `--parallelismlimit`—This refers to how much manifest generation can be done concurrently. The default value is 0, which means there is no limit at all, and this increases the chance of getting an OOM exception that will restart the container. That is why it is useful to track OOM restarts (we looked at this in *Chapter 3*, *Operating Argo CD*, in the *Enabling observability* section), and once you see you have quite a few of those, you can introduce a limit. There is also an environment variable you can use instead of flag: `ARGOCD_REPO_SERVER_PARALLELISM_LIMIT`.

- `--repo-cache-expiration`—This is the duration for which manifests are cached, and the default value is 24 hours. Normally, you should not modify this value, but depending on your deploy frequency, you can go even higher. Setting a smaller value will reduce the benefit of caching because it makes Argo CD recalculate the manifests, and that's an expensive operation.

- `ARGOCD_GIT_ATTEMPTS_COUNT` environment variable—This is used to specify how many times we try to transform the tag, branch reference, or HEAD of the Git repository default branch into the corresponding commit SHA. This operation is done many times, and sometimes it can fail, which leads to failed synchronizations. The default value is 1, which means no retries, so it would be safe to set a value bigger than 1.

- `ARGOCD_EXEC_TIMEOUT` environment variable—When manifest generation for Helm or Kustomize happens, it is likely that there will be a need to download something—either a Helm chart from a registry or a remote base for Kustomize. This adds to the time needed to generate manifests, which can also take some time—especially for big Helm charts—and with this variable, we can control the timeout for the manifest generation inside the repository server. The default value is 90 seconds, but if you see a lot of failed reconciliations, you can increase it. Also, please note that this is connected to the `--repo-server-timeout-seconds` flag from the application controller, and they should have similar values.

If you need to use different values for Helm parameters , we need to specify them under `repoServer.extraArgs` for the flags and under `env` for the environment variables. The updated section of the `values.yaml` file would look like this (a complete override can be found in the official repository at `https://github.com/PacktPublishing/ArgoCD-in-Practice`, in the `ch7/performance-improvement/helm-installation/override-values.yaml` file):

```
repoServer:
  extraArgs:
```

```
    - --parallelismlimit=5
    - --repo-cache-expiration=24
  env:
    - name: "ARGOCD_GIT_ATTEMPTS_COUNT"
      value: "3"
    - name: "ARGOCD_EXEC_TIMEOUT"
      value: "300s"
```

The main server application does not have any flags that could affect performance because the repository server and application controller do the heavy lifting. The server has the main task of exposing an API that is consumed by the web UI, CLI, and any other custom tools that we might build (for example, by using `curl`).

Summary

In this chapter, we went through some things that can happen with Argo CD that you would not expect, from an issue that can occur at installation to how restarts can be performed safely and how to stick to a specific version of a templating engine. In the end, we took a closer look at some of the most important configuration parameters and environment variables that could affect and improve the performance of two of the most important components of Argo CD: the repository server and the application controller.

In the next chapter, we will take a close look into how we can perform different checks and validations on the YAML file we are pushing to Argo CD in order to catch as many issues as possible before applying the manifests to production.

Further reading

For more information, refer to the following resources:

- Argo CD and Kustomize: `https://redhat-scholars.github.io/argocd-tutorial/argocd-tutorial/03-kustomize.html`

- Kubernetes `emptyDir` volume type: `https://kubernetes.io/docs/concepts/storage/volumes/#emptydir`

- More and better explanations on how `mountPath` and `subPath` work together: `https://stackoverflow.com/questions/65399714/what-is-the-difference-between-subpath-and-mountpath-in-kubernetes/65399827`

- General troubleshooting techniques of Kubernetes applications: `https://kubernetes.io/docs/tasks/debug-application-cluster/debug-application/`

- Performance tuning is an ongoing thing—watch this thread for more details and developments: `https://github.com/argoproj/argo-cd/issues/3282`

8

YAML and Kubernetes Manifests

We can't talk about GitOps and Argo CD and not have a chapter dedicated to **YAML Ain't Markup Language** (**YAML**). We wrote a lot of YAML in all the chapters so far, and I expect you will write a lot more if you start using Argo CD, so we are going to check some ways to statically analyze it. First, we will take a close look at the most common templating engines, Helm and Kustomize, and how we can use them in order to generate the final manifests our GitOps engine is going to apply. Then, we will look at a tool that can validate the manifests we will be creating against the Kubernetes schema. After this, we will check the most common practices to enforce on the manifests, which helps us to introduce stability and predictability into the system. And we will finish the chapter by introducing one of the most interesting tools to use in pipelines in order to perform extended checks over YAML—`conftest`, which allows you to write your own rules in a language called Rego.

The main topics we are going to cover in this chapter are listed here:

- Working with templating options
- Exploring types of validation
- Validating a Kubernetes schema
- Enforcing best practices for your manifests
- Performing extended checks with `conftest`

Technical requirements

In this chapter, we will concentrate on the steps we can take before the YAML is being merged to the GitOps repository. Once that code reaches the main branch it will get applied to the cluster by Argo CD, so we will take a look at the possible validations we can perform prior to the merge. While not mandatory to have a running installation of Argo CD, it will still be good to have one, so you can check the installations of the applications we will create in the end. We will need to have Helm (`https://helm.sh/docs/intro/install/`) and Kustomize (`https://kubectl.docs.kubernetes.io/installation/kustomize/`) installed. We will also use Docker (`https://docs.docker.com/get-docker/`) to run containers for all the tools we will use in the demos. So, we will not install them; instead, we will be using their container images. All the code we will be writing can be found at our official repository `https://github.com/PacktPublishing/ArgoCD-in-Practice`, in the `ch08` folder.

Working with templating options

We want to take a look at the main YAML templating options, Helm and Kustomize, and how you can get the best out of them when used with Argo CD. We are not going to introduce how these work as we expect that you have some knowledge of these tools. If you are not familiar with them, please follow their official guides—for Helm, we have `https://helm.sh/docs/intro/quickstart/`, and for Kustomize, there is `https://kubectl.docs.kubernetes.io/guides/`. Instead, we will be focusing on how you can generate manifests from templates in the same way as done by Argo CD.

Helm

Helm is probably the most used templating option for Kubernetes manifests. It is very popular and widely adopted, so you will probably deploy most of your applications using Helm charts. The easiest way to start installing Helm charts into a cluster is to use the native declarative support of Argo CD applications. We can see how we will be able to deploy a Traefik chart with this approach (`https://github.com/traefik/traefik-helm-chart`). Traefik plays the role of an ingress controller, allowing us to handle incoming traffic in our Kubernetes clusters. It will make the connection between an external endpoint and an internal Service and also enables us to define all sorts of middleware components.

So, we want to deploy a Traefik chart, version `9.14.3`, and for starters, we will have some parameters overridden from their default values: for example, we want three replicas instead of one; we want to enable a **PodDisruptionBudget** (**PDB**), which makes it possible to define how many pods can be unavailable or available in case of an unexpected event (a cluster upgrade can be an example of such an event); we want to have the logs level on `INFO` and to also enable the access logs.

You can find the Application in the `ch08/Helm/traefik-application.yaml` folder from the official repository of this book, and you can apply it to an existing installation like this:

```
kapply -f ch08/Helm/traefik-application.yaml -n argocd
```

This is what the Application should look like:

```
apiVersion: argoproj.io/v1alpha1
kind: Application
metadata:
 name: traefik
 namespace: argocd
spec:
 project: default
 source:
   chart: traefik
   repoURL: https://helm.traefik.io/traefik
   targetRevision: 9.14.3
   helm:
     parameters:
       - name: image.tag
         value: "2.4.2"
       - name: deployment.replicas
         value: "3"
       - name: podDisruptionBudget.enabled
         value: "true"
       - name: logs.general.level
         value: "INFO"
       - name: logs.access.enabled
         value: "true"
 destination:
   server: https://kubernetes.default.svc
   namespace: traefik
 syncPolicy:
   automated: {}
   syncOptions:
     - CreateNamespace=true
```

These are just a few examples of parameters that we can override. In a real installation, we will have many more that will be different than the default ones. It's likely we will need the Service to be of type LoadBalancer and then control different settings with the help of annotations. Then, we will need to set resources for the containers, maybe some additional Pod annotations if we are using a Service mesh, and so on. The problem that I am trying to highlight is that for once, the Application file could become really big because of many parameters to overwrite, but also we are making a tight coupling between the Argo CD Application details and the Helm chart details. It would be much better if we could separate these two parts—the Application definition and the Helm chart.

And that would be possible if we used the Helm chart we want to deploy as a subchart. Alternatively, you might also sometimes come across the term *umbrella chart*, which means we will be using a chart that defines a dependency on the needed chart, thus becoming a subchart, or the main chart becoming an umbrella. In this case, we will have a folder called traefik-umbrella, and inside the folder, we need at least two files to define the subchart. The first file, Chart.yaml, should look like this:

```
name: traefik-umbrella
apiVersion: v2
version: 0.1.0
dependencies:
- name: traefik
  version: 9.14.3
  repository: "https://helm.traefik.io/traefik"
  alias: traefik
```

And then, the values.yaml file will only contain our overwrites, as illustrated in the following code snippet:

```
traefik:
 image:
   tag: "2.4.2"
 deployment:
   replicas: 3
 podDisruptionBudget:
   enabled: true
 logs:
   general:
     level: "INFO"
   access:
     enabled: "true"
```

Next, we create an Application that points to the folder where these two files are located. Argo CD notices the Chart.yaml file and realizes that it is about a Helm chart. Initially, it will download all the dependencies using the helm dependencies update command, after which it will call helm template in order to generate the manifests that will get applied to the cluster. This process can sometimes lead to an error (for example, a timeout while downloading the dependencies or a dependency version that doesn't exist), and this will only be visible in the Argo CD Application state. So, it would be good if we could catch such errors before the Helm chart is processed by Argo CD, such as in a pipeline that could run before merging the YAML changes to the main branch.

Following the subchart approach, when we do an upgrade, we need to modify the version in Chart. yaml—for example, in our case, we can upgrade from 9.14.3 to 9.15.1. The update looks like a simple one, but on the chart itself, there could be many changes made, resources added, default values modified, and other things that could affect the application behavior. So, it would be good if we could get an indication of what types of changes the new version introduces.

> **Note – Argo CD updates CRDs automatically**
>
> If you are familiar with how Helm 3 handles **Custom Resource Definitions** (**CRDs**), you would know that they are not updated automatically (only the initial creation is done automatically). Instead, you need to manually handle the update: https://helm.sh/docs/chart_best_practices/custom_resource_definitions/. But Argo CD will not treat them in a special way, and they will be updated as any other resource. That's because Helm is only used to generate manifests, after which they are applied with kubectl.

We can do all this: catch any errors regarding the Helm YAML templating and check for the introduced changes with a simple script that can be run in our pipeline (the script can be found in the official repository, in the ch08/Helm/validate-template.sh file). We fetch the Traefik subchart, and then we generate all the manifests to an output directory. Then, assuming we are doing this from a feature branch and not the main one, as it would be a good idea to run the script part of a **pull request** (**PR**), we check out main and again generate templates in a different output folder this time. In the end, we call diff to get the differences between the two folders where we generated the Kubernetes manifests and we do this with a || true suffix because diff returns an error exit code if differences are found, while for us it is not an exception. The following snippet shows all the code involved:

```
helm dependency update traefik-umbrella
helm template traefik-umbrella –include-crds --values traefik-
umbrella/values.yaml --namespace traefik --output-dir out
git checkout main
helm dependency update traefik-umbrella
helm template traefik-umbrella –include-crds --values traefik-
umbrella /values.yaml --namespace traefik --output-dir out-
default-branch
diff -r out-default-branch out || true
```

If you only want to verify that your changes will not give an error when applied by Argo CD, you will need to run just the first two commands. And if the templates give an error that is not easy to understand, there is always the `--debug` flag that can be used in the `helm template` command (more on this at `https://helm.sh/docs/chart_template_guide/debugging/`).

Because of Helm's popularity, you will end up installing a lot of Helm charts with Argo CD. The main applications out there already have a Helm chart created that you can easily import and use in your clusters. So, it will be worth investing the extra time into creating a pipeline for dealing with Helm charts—the sooner you do it, the more you can benefit from it, and you will throw fewer issues at Argo CD because you will catch them beforehand with the validations that run in your pipeline.

Kustomize

Kustomize, while not as popular as Helm, is still an option that application developers like to use. Argo CD initially only offered simple manifests for Kustomize to install, and after a while, a Helm chart was created. I know many people who were early adopters of Argo CD that still deploy it with Kustomize, which had quite significant advantages as compared to Helm version 2, but this is no longer the case after Helm version 3 was released.

When Argo CD looks into the Git repository, it realizes that it is about Kustomize if it finds the `kustomization.yaml` file. If the file exists in the repository, then it will try to generate manifests using the `kustomize build` command (`https://github.com/kubernetes-sigs/kustomize#1-make-a-kustomization-file`), and—as usual—when the manifests are ready, they will be applied with `kubectl apply` to the destination cluster. So, we can apply the same logic as we did with Helm templating to generate original manifests from the main branch and updated manifests from the current working branch, and then compare them with the `diff` command and display the result for the user to see. As there are fewer applications deployed these days with Kustomize, and also because it is pretty similar to what we did with Helm templates, we are not going to do any demo of this.

Generating the manifests that Argo CD will apply is a good place to start to have some prior validation of the Kubernetes resources before handing them over to our GitOps operator. If there is an issue, we can find it in advance, thus avoiding the need to check the application status after the sync, to see if everything was alright on the side of Argo CD. This way, we can reduce the number of errors thrown after we merge our changes by shifting left and running these checks on every commit. Next, we will go through the types of validations we can perform for our manifests, from the ones that make checks on the YAML structure to the ones that understand Kubernetes resources.

Exploring types of validation

There is a lot of YAML in the Kubernetes world, and we have seen how we can use templating options around it in order to improve it. We usually start with small manifests, and then we keep adding resources to our Helm chart and then more environment variables, an `init` container, sidecars, and so on. In the end, we will have a big Helm chart or Kustomize manifests that we will apply to the cluster.

So, it would be a good idea to validate them before applying them, in order to catch as many issues as possible before they turn into bigger issues in our clusters that can only be fixed with manual intervention.

We can start with the simplest linting of all, and that's of YAML files' structure, without looking at the syntax—so, making abstraction that it contains Kubernetes manifests. A good tool to verify that your YAML content doesn't have any cosmetic issues—such as too big lines or trailing spaces, or the more problematic indentation issues—is `yamllint`: `https://github.com/adrienverge/yamllint`. This embeds a set of rules that check different things on your YAML files, and it also gives you the possibility to extend it by creating your own rules.

It is easy to start using it in a pipeline because a Docker image is provided, and it can be found in both Docker Hub (`https://hub.docker.com/r/pipelinecomponents/yamllint`) and GitLab registry (`https://gitlab.com/pipeline-components/yamllint`). It has extensive documentation (`https://yamllint.readthedocs.io/en/stable/index.html`) and it provides many configuration options, such as disabling rules or modifying default values (for example, the max length of a YAML line can be configured).

We should not expect `yamllint` to surface big issues from our manifests; in the end, it checks that the structure is in order, but if we follow its rules, we will end up having YAML manifests that are easy to read and understand.

Once our YAML files look neat, we can go on and introduce more tools we can use for validating our manifests. Because Helm is the most used option for templating, there are also tools created especially for testing Helm charts. The most important one is the official `chart-testing` project or `ct` (`https://github.com/helm/chart-testing`). Besides validating a Helm chart, it can also install it and perform live tests on it (which under the hood will run the `helm install` and `helm test` commands). We will look closer at how we can do the linting part as that can give us fast feedback without actually performing an installation, so there is no need for a real Kubernetes cluster.

The command for static validation is `ct lint` (`https://github.com/helm/chart-testing/blob/main/doc/ct_lint.md`), and under the hood, it will run `helm lint` and many other actions, such as chart version validations and confirms that you bumped it in your feature branch compared to the default branch, and even checks that chart maintainers are part of your GitLab top group or GitHub organization.

It does support linting with multiple `values.yaml` files that you can define in the `ci` folder, directly under our chart, making it easy to test different logic based on different inputs provided in the YAML file.

Another advantage is that the `ct` project embeds the `yamllint` tool we described before and even another one called `yamale` (`https://github.com/23andMe/Yamale`), also used for YAML structure linting, so there is no need to use them separately—they will be run once with the `ct lint` command.

There is a container image provided, which makes it easy to run it as part of **continuous integration** (**CI**) pipelines or locally—for example, as part of a Git pre-commit hook (`https://quay.io/repository/helmpack/chart-testing`).

Besides these tools used to validate YAML and Helm charts, we have the ones that understand Kubernetes resources, including all the relations and constraints that can be defined for them. So far, there are three types of such tools, as outlined here:

- Those that can validate the manifests according to a Kubernetes schema version so that we can run checks before upgrading to a newer version.

- Those that verify the manifests follow a (fixed) list of best practices.

- Those that allow you to define your own rules for validation. They are harder to use as we need to learn how to build those new rules, but they give you big flexibility this way.

Next, we will explore all the three types of tools we described, and we will start by looking at how we can prepare for the schema updates introduced by new Kubernetes versions.

Next, we are going to explore a tool that looks more promising in each of these three categories. We will check out `kubeconform` to perform **application programming interface** (**API**) validations, then `kube-linter` for making sure we follow best practices, and in the end, we will take a close look at `conftest`, a powerful tool that allows us to build our own rules in the Rego language.

Validating a Kubernetes schema

New Kubernetes versions can come with API version deprecations and removals. For example, in release `1.16` for `CustomResourceDefinition`, the `apiextensions.k8s.io/v1` version was introduced, while in `apiextensions.k8s.io/v1beta1` it was deprecated, and later, in version `1.22`, it was completely removed. So, if you used `apiextensions.k8s.io/v1beta1` for `CustomResourceDefinitions` starting with Kubernetes version `1.16`, you would get a deprecation warning, while if you used it with version `1.22`, you would get an error because the version doesn't exist anymore.

Usually, the problem is not only about applying older and unsupported API versions to a Kubernetes cluster; instead, it may be more likely that you have a deprecated version already installed with some applications, after which you upgrade the cluster to a version where the API is completely removed. Normally, you should catch it while you upgrade the development or testing clusters, but there is always a slight chance of missing the error and ending up with it in your production cluster.

So, a better way would be if we could verify the manifests we want to apply against specific versions of Kubernetes before we actually apply them, and we can achieve this in a few ways. First, we can run `kubectl apply` with the `--dry-run=server` option, which will take the manifests and send them to the server for validation, but the changes will not be saved. We also have a corresponding `--dry-run` flag that we can use for the `helm install` command. This approach works very well, but the problem is that you need a Kubernetes cluster with a specific version up and running. For example, if you have now in production Kubernetes version `20`, and you want to validate your manifests for versions `21` and `22`, you would need one cluster for each version. If you start them on demand, it takes time—at least a couple of minutes nowadays—so, that's not fast enough feedback that you can wait for every commit. If you keep all three versions up and running all the time, that will mean you would need to pay a higher cost to the cloud provider for the infrastructure, and it also means additional effort to support them.

Another way to validate your manifests would be to use the schema of the Kubernetes API. This is described in the OpenAPI format (`https://kubernetes.io/docs/concepts/overview/kubernetes-api/`), with the schema file being in the Kubernetes Git repository, which can be transformed into a **JavaScript Object Notation (JSON)** schema that can then be used to validate our YAML manifests (as YAML is a superset of JSON). And we have tools that do all these transformations and validations automatically, so you don't need to deal with them, just pass via parameters the Kubernetes version to use for validation.

The first tool that made this possible was `kubeval` (`https://github.com/instrumenta/kubeval`), but lately, it doesn't look as though this project is under active development anymore. Their latest version, `v0.16.1`, is around a year old, and there haven't been any more accepted PRs for many months now. This opened the way for other tools to accomplish schema validation, and among them, it seems that `kubeconform` (`https://github.com/yannh/kubeconform`) is the most advanced one.

So, let's see how we can validate the manifests we generated for Traefik with different schema versions. We will not need to install `kubeconform` as we will use the container image to run the tool. The latest version right now is `v0.4.12`, and I will be using the `amd64` version of the container image as that's the most common (and also my) **central processing unit (CPU)** architecture. You also have **Advanced RISC Machine (ARM)** container versions; you can find them all here: `https://github.com/yannh/kubeconform/pkgs/container/kubeconform`. Going back to the Traefik Helm chart, we will first need to generate manifests, and then we will run `kubeconform` validations for versions `21` and `22`.

CRDs are a little bit different than other types in Kubernetes; their schema is based on OpenAPI v3 and not v2 (`https://kubernetes.io/docs/tasks/extend-kubernetes/custom-resources/custom-resource-definitions`). That's why `kubeconform` (and also `kubeval`) can't handle validations for them very well because their tools are failing to import the OpenAPI v3 schema. So, if you run `kubeconform` directly to validate CRDs, you will get an error that the schema was not found. We will see this in action for our Traefik Helm chart; the commands can be found in the `ch08/kubeconform/validate-schema.sh` file. We will run `kubeconform` with its container image, passing the path to where the manifests were generated as a volume to the container (this will work only for Linux distributions and macOS. For Windows, please replace $(pwd)/out with the full path to the /out folder). The code is illustrated here:

```
helm dependency update traefik-umbrella
helm template traefik-umbrella –include-crds --values traefik-
umbrella/values.yaml --namespace traefik --output-dir out
docker run -v $(pwd)/out:/templates ghcr.io/yannh/
kubeconform:v0.4.12-amd64 -kubernetes-version 1.21.0 /templates
```

The command will give us several warnings, such as these:

```
/templates/traefik-umbrella/charts/traefik/crds/
traefikservices.yaml - CustomResourceDefinition
traefikservices.traefik.containo.us failed validation: could
not find schema for CustomResourceDefinition
/templates/traefik-umbrella/charts/traefik/templates/dashboard-
hook-ingressroute.yaml - IngressRoute RELEASE-NAME-traefik-
dashboard failed validation: could not find schema for
IngressRoute
```

This means we are missing definitions for these types, and we have two options to choose from— either we skip validating them or we provide the missing definitions. In our case, we would still like to validate CRDs, but we can skip IngressRoute as that is something introduced by the Traefik chart and not related to a Kubernetes version. For skipping this, we have the -skip flag. For the CRD definition, there are already generated ones under https://jenkins-x.github.io/ jenkins-x-schemas/. For validating version 1.21, we will use apiextensions.k8s.io/ v1beta1 as it is still available, while for version 1.22, we will use apiextensions.k8s.io/ v1 as the previous one was removed. For version 1.21, this will be the command:

```
docker run -v $(pwd)/out:/templates ghcr.io/yannh/
kubeconform:v0.4.12-amd64 -kubernetes-version 1.21.0 -skip
IngressRoute -schema-location default -schema-location
'https://jenkins-x.github.io/jenkins-x-schemas/apiextensions.
k8s.io/v1beta1/customresourcedefinition.json' /templates
```

And we will not get any error, which is the expected behavior. Then, the command for version 1.22 will look like this:

```
docker run -v $(pwd)/out:/templates ghcr.io/yannh/
kubeconform:v0.4.12-amd64 -kubernetes-version 1.22.0 -skip
IngressRoute -schema-location default -schema-location
'https://jenkins-x.github.io/jenkins-x-schemas/apiextensions.
k8s.io/v1/customresourcedefinition.json' /templates
```

And here, we get a list of errors for every CRD similar to this item, which is what we were expecting because of the differences between `v1beta1` and `v1` of `apiextensions.k8s.io`:

```
/templates/traefik-umbrella/charts/traefik/crds/ingressroute.
yaml - CustomResourceDefinition ingressroutes.traefik.containo.
us is invalid: For field spec: Additional property version is
not allowed
```

> **Note – Kubernetes Is Moving To OpenAPI v3**
>
> Starting with version `1.23`, Kubernetes is starting to migrate to OpenAPI v3, for now only in an alpha state. The main advantage for Kubernetes schema validating tools is that now the schema is published in Git in v3 format and that there is no need to transform CRDs from v3 to v2 to be validated: `https://github.com/kubernetes/enhancements/tree/master/keps/sig-api-machinery/2896-openapi-v3`.

There is an active discussion in `kubeconform` GitHub issues on how to make the validations of CRDs easier. You can follow it here: `https://github.com/yannh/kubeconform/issues/51`.

Validating the schema of each version in the CI part before applying the changes allows us to shift left and catch any possible issues as fast as possible. This is one type of validation that we can do by statically analyzing the YAML manifests, but there are others we can perform. So, next, we are going to see a tool that checks the Kubernetes manifests for a series of best practices.

Enforcing best practices for your manifests

When you go live with your applications on Kubernetes, you want things to go as smoothly as possible. In order to achieve that, you need to get prepared in advance and be up to date with all the important items when deploying on Kubernetes, from making sure that all Pods have defined readiness and liveness probes (`https://kubernetes.io/docs/tasks/configure-pod-container/configure-liveness-readiness-startup-probes/`) and that the containers have memory and CPU limits and requests (`https://kubernetes.io/docs/concepts/configuration/manage-resources-containers/`), up to checking that the latest tag is not used for container images (`https://kubernetes.io/docs/concepts/containers/images/`). And then, you need to make sure that the product teams are following these practices when building their Helm charts. Most likely, your organization plans to add more and more microservices to production, and also, you find new and more complex practices that the developers need to be aware of. So, ideally, it would be great if you could automate such validations and add this as a gate in your CI pipelines.

The good thing is that there are tools that can accomplish such checks, and two of the most known ones are `kube-score` (`https://github.com/zegl/kube-score`) and `kube-linter` (`https://github.com/stackrox/kube-linter`). These are pretty similar in terms of the end result, and they are both easy to add to CI pipelines as they provide container images. `kube-score` is the older project, so it might be used by more people, but `kube-linter` has more GitHub stars (2k for `kube-linter`, compared to 1.8k for `kube-score` as of August 2022). In the end, I will go with `kube-score` for our demo as I like the fact that it provides container images that also bundle Helm or Kustomize. This will make our tasks a little easier for generating manifests to scan, and the fact is, right now, `kube-linter` is declared to be in an alpha stage. That may change in the future, so if you have to pick one or the other, you should make your own thorough analysis.

The checks that `kube-score` performs are categorized into critical and warning ones, and the tool will automatically return exit code 1 if it finds any critical issues. It has a flag to also exit with code 1 if only warnings have been found: `--exit-one-on-warning`. It also has a list of optional validations that could be performed that you would need to enable with the `--enable-optional-test` flag, followed by the name of the test. There is also the possibility to ignore a check that is performed by default with the `--ignore-test` flag followed by its name, so if you know that you are not ready yet for some of them, you can stop checking them for a period.

Next, let's see how we can perform some checks for our Traefik installation and how we can address some of the findings, skip some of them, or enable some of the optional ones. We will do this using the container image of the `1.14.0` version: `zegl/kube-score:v1.14.0`. `kube-score` also provides container images with Helm 3 or Kustomize bundled, so if your **continuous integration/continuous deployment (CI/CD)** engine supports running containers directly (such as GitLab CI/CD), you can use those and perform the manifest generation inside the container. In our case, we will create a script that uses the container image as we would run it in a **virtual machine (VM)**. We will perform a check without any parameters, based on the default values. For a complete list of which tests are enabled by default and which are optional, you can check out the following link: `https://github.com/zegl/kube-score/blob/master/README_CHECKS.md`.

It is a little harder to get the `docker run` command running this time because `kube-score` expects the exact YAML filename as input, so the part with globs from the path (`/**/*.yaml`) is handled by the shell. To make it work in our case, the relative paths inside the container and on the machine have to be the same. For this, I set the working directory in the container to `/` (with the `-w` flag), while on the machine we need to be in the directory where the `out` folder is. The script can be found in the official repository, in the file located at `ch08/kube-score/enforcing-best-practices.sh`. The code is illustrated here:

```
helm dependency update traefik-umbrella
helm template traefik-umbrella –include-crds --values traefik-
umbrella/values.yaml --namespace traefik --output-dir out
docker run -v $(pwd)/out:/out -w / zegl/kube-score:v1.14.0
score out/traefik-umbrella/charts/traefik/**/*.yaml
```

The command results in a series of critical issues and warnings, most of them being on the `Deployment` manifest. We will post only a few of them here:

```
apps/v1/Deployment RELEASE-NAME-traefik
    [CRITICAL] Container Image Pull Policy
        · RELEASE-NAME-traefik -> ImagePullPolicy is not set to
Always
            It's recommended to always set the ImagePullPolicy
to Always, to
            make sure that the imagePullSecrets are always
correct, and to
            always get the image you want.
. . . . . . . . . . . . . . . . . . . . . . . . . . . . . . . . . . . . .
    [CRITICAL] Container Resources
        · RELEASE-NAME-traefik -> CPU limit is not set
            Resource limits are recommended to avoid resource
DDOS. Set
            resources.limits.cpu
        · RELEASE-NAME-traefik -> Memory limit is not set
            Resource limits are recommended to avoid resource
DDOS. Set
            resources.limits.memory
        · RELEASE-NAME-traefik -> CPU request is not set
            Resource requests are recommended to make sure that
the application
            can start and run without crashing. Set resources.
requests.cpu
        · RELEASE-NAME-traefik -> Memory request is not set
            Resource requests are recommended to make sure that
the application
            can start and run without crashing. Set resources.
requests.memory
    [WARNING] Deployment has host PodAntiAffinity
        · Deployment does not have a host podAntiAffinity set
            It's recommended to set a podAntiAffinity that
stops multiple pods
            from a deployment from being scheduled on the same
node. This
            increases availability in case the node becomes
unavailable.
```

Here, we can also see what the command would look like if we would like to skip the check for the container image pull policy to be `Always` and enable the optional test for validating that all your containers have a `seccomp` policy configured, used to restrict a container's **system calls (syscalls)**:

```
docker run -v $(pwd)/out:/out -w / zegl/kube-score:v1.14.0
score --ignore-test container-image-pull-policy --enable-
optional-test container-seccomp-profile out/traefik-umbrella/
charts/traefik/**/*.yaml
```

`kube-score` has performed more than 30 checks today if we count both the default and optional ones, but even with this number, it is not a comprehensive check of your manifests. It does try to give you some insights into things that can possibly go wrong, and because of this, it can also be considered a good learning resource. When you start checking out more details on the tests it is performing, you can find a lot of relevant information on how to make your cluster stable and reliable.

Performing extended checks with conftest

Open Policy Agent (**OPA**) (`https://www.openpolicyagent.org`) is an engine that can validate objects prior to performing a change on them. Its main advantage lies in the fact that it doesn't come with a predefined list of checks; instead, it supports extensible policies as they are based on rules created in the Rego language (`https://www.openpolicyagent.org/docs/latest/policy-language/`). You might have heard of OPA in conjunction with Kubernetes: that it can be used like an admission controller (a part usually handled by the Gatekeeper project: `https://github.com/open-policy-agent/gatekeeper`) in order to add a pre-validation of the objects you want to apply in a cluster. OPA is really successful at adding policy-as-code checks for Kubernetes, but it is more than that: it is an engine that can be run almost everywhere we have a runtime, including in our CI/CD pipelines.

For Kubernetes, you can create your own custom rules to be enforced by OPA. For example, you can have a policy that says every namespace needs to have a label that specifies the team that owns it, or a policy that states that every container you deploy on the cluster needs to be from a pre-approved list of registries, or even registry users or organizations. But it would be even better if we could run such policies in the pipeline without actually applying them on the cluster so that we would get feedback much faster from our CI/CD system before we could merge this and hand over the changes to Argo CD, and there is an open source tool that can accomplish that with an embedded OPA engine: `conftest` (`https://www.conftest.dev`).

`conftest` allows us to create policies in Rego and run them in our pipelines, having the OPA engine validate the manifests. The rules that we create for `conftest` are not exactly the same as the ones we use in the Kubernetes admission controller, but they are similar, and it is easy to adapt them to one side or the other.

We will go through an example to check if our container image either comes from our private registry or if it can be from Docker Hub but needs to be an official image (`https://docs.docker.com/`

`docker-hub/official_images/`), which are much safer to use. Let's say that our private registry is in Google Cloud (`https://cloud.google.com/container-registry`) and that all images should start with `eu.gcr.io/_my_company_/`, while the Docker Hub official images are the ones that don't have any user, so their format should be without any additional / sign, such as `traefik:2.4.2` and not `zegl/kube-score:v1.14.0`.

Rego is not an easy language that can be picked up in a few hours, but it is also not too complicated once you get beyond the basics. Here are a few things to consider when you start writing or reading Rego. This is not a comprehensive list, but something to get you started:

- Every statement returns `true` or `false` (statements such as `assignment` always return true).
- `input` is a special keyword and it is the root of the JSON (or YAML) object it analyzes.
- Rules are a set of instructions that allow us to take a decision, while functions are similar to functions from other programming languages—they take an input and return a value.
- On all the expressions inside a function or rule AND is applied, which means they all need to be `true` for the end result to be `true`.
- A function or rule can be defined multiple times, and on the result of all of them OR will be applied.
- With `conftest`, we are allowed to use only special rules, such as `deny`, `violation`, and `warn`. If the result of the rule is false then `conftest` will exit with an error, and this way, we can stop the pipeline.

You will find a lot of resources about Rego online and how you can use it with OPA or `conftest`, and I recommend you also check out this video, which is an OPA deep dive from 2018 (a little older, but still relevant to how Rego works):

`https://www.youtube.com/watch?v=4mBJSIhs2xQ`

In our case, we would define two functions both named `valid_container_registry`, the first checking if the registry used is Docker Hub with an official image, meaning there is no / sign in the image name, while the second function will verify that the first and second values for a split of the image by / will be `eu.gcr.io` and `_my_company_`. The code can also be found in the official repository in the `ch08/conftest` folder. The policy that we defined and that you can see next is in the `policy/deployment.rego` file, because `conftest` expects by default all policies to be under the `policy` folder, while the resources we analyzed are under the `manifests` folder. This is what the functions and the `deny` rule look like:

```
package main

deny[msg] {
    input.kind == "Deployment"
```

```
    container_registry_image := input.spec.template.spec.
containers[_].image
    output = split(container_registry_image, "/")
    not valid_container_registry(output)
    msg = "invalid container registry in the deployment"
}
valid_container_registry(imageArr) = true {
    count(imageArr) == 1
}
valid_container_registry(imageArr) = true {
    count(imageArr) > 2
    imageArr[0] == "eu.gcr.io"
    imageArr[1] == "_my_company_"
}
```

And you can run `conftest` with this command, using its container image and attaching a volume to the container that has the policy and the manifests to analyze:

```
docker run -v $(pwd):/project openpolicyagent/conftest:v0.30.0
test manifests/
```

The result should be similar to the following:

```
2 tests, 2 passed, 0 warnings, 0 failures, 0 exceptions
```

The image that we are using in the analyzed Deployment, `traefik:2.4.2`, is an official Docker Hub one, so that's why our checks are passing. Feel free to modify the manifests for test purposes and run `conftest` again to see how it fails.

Writing policies is not an easy task, and the community started gathering them in repositories in order to share them. You have projects such as `https://github.com/open-policy-agent/library` or `https://github.com/redhat-cop/rego-policies`, and I also want to share one repository that tries to gather all the OPA/Rego/`conftest` resources together, such as documentation, articles, books, additional tools, or package libraries: `https://github.com/anderseknert/awesome-opa`.

`conftest` is much more than Kubernetes manifests' validation. Besides YAML/JSON, it can do checks for many other declarative languages. It currently supports **HashiCorp Configuration Language (HCL)** and HCL2, so we can write policies for Terraform on infrastructure provisioning, Dockerfiles to check container creation, and others such as **initialization (INI)**, **Tom's Obvious Minimal Language (TOML)**, **Extensible Markup Language (XML)**, Jsonnet, and so on, which means it is worth checking it and trying it out as it has a lot of potential for defining gates in many types of pipelines.

Summary

In this chapter, we went through some of the options we have to statically analyze Kubernetes YAML manifests. We saw how we can generate manifests from templating engines such as Helm or Kustomize, and then we checked some tools that can perform several types of jobs: `kubeconform` will validate your manifests against the OpenAPI Kubernetes schema, `kube-score` will check that you follow a predefined list of best practices, while `conftest` can do everything because it allows you to define your own rules and policies for the manifests to follow. All these validations can be easily added to your CI pipeline, and we have seen examples of how to use them directly with their container images.

In the next chapter, we will take a close look at what the future might bring for Argo CD and how it can be used to democratize and standardize GitOps with GitOps Engine, an innovative project built with the help of other organizations from the community that is already seeing some good adoption rates in the industry.

Further reading

For more information, refer to the following resources:

- More options and tools for YAML validating: `https://learnk8s.io/validating-kubernetes-yaml`

- Kubernetes schema validation: `https://opensource.com/article/21/7/kubernetes-schema-validation`

- Detailed view of the checks `kube-score` performs: `https://semaphoreci.com/blog/kubernetes-deployments`

- OPA architecture: `https://www.openpolicyagent.org/docs/latest/kubernetes-introduction/`

- Use `conftest` for your Terraform pipelines: `https://dev.to/lucassha/don-t-let-your-terraform-go-rogue-with-conftest-and-the-open-policy-agent-233b`

Future and Conclusion

In this chapter, we will describe the democratization of GitOps under common practices. We will go through GitOps Engine and the `kubernetes-sigs/cli-utils` libraries and understand how various GitOps operators out there will be able to reuse a common set of GitOps features.

Lastly, we will go through the learnings of this book and what is next after you have finished reading the *Argo CD in Practice* book.

The main topics we will cover are as follows:

- Democratizing GitOps
- What is GitOps Engine?
- What is `kubernetes-sigs/cli-utils`?
- Wrap up

Technical requirements

For this chapter, we assume that you have already familiarized yourself with the `kubectl` commands. Additionally, you will need the following:

- `git-sync`: `https://github.com/kubernetes/git-sync`

The code can be found at `https://github.com/PacktPublishing/ArgoCD-in-Practice/tree/main/ch09`.

Democratizing GitOps

In this section, we will go through the various Go libraries that have been created in order to provide a common set of core features for the different GitOps operators in the industry. We will go through the following items:

- What is GitOps Engine?
- What is `cli-utils` for Kubernetes SIGs?

What is GitOps Engine?

The Argo CD team announced the implementation of GitOps Engine a few years back. Different GitOps operators in the industry address several different use cases but they are all mostly a similar set of core features. Based on this, the Argo CD team had an idea to build a reusable library that implements the core GitOps principles. The core features of GitOps Engine are the following:

- A cache
- Diffing and reconciliation
- Health assessment
- Synchronization
- Engine

Argo CD is already updated and uses GitOps Engine and, in parallel, still keeps the enterprise features so it can bring the GitOps power to enterprise organizations too. The features that Argo CD delivers to GitOps as a service to large enterprises and that are not part of GitOps Engine are the following:

- Single sign-on integration
- Role-based access control
- Custom resource definitions abstractions such as `Application` and `Project`, which we have already discussed in earlier chapters

All these features combined with GitOps Engine enable engineering teams to adopt GitOps much easier. Let's take a look at the GitOps agent, a real example using GitOps Engine that has been built by the Argo team.

The GitOps agent

The Argo team, based on GitOps Engine, created a basic GitOps operator that leverages the GitOps Engine packages we described to support simple GitOps operations use cases. So, the agent doesn't include any of the features of Argo CD, such as multi-repositories, single sign-on, or multi-tenancy.

The main idea around the GitOps agent is to inspire us to use the reusable library of GitOps Engine and share feedback with the Argo team or contribute to GitOps Engine with extra features.

Usage

The core difference between the GitOps agent and Argo CD is that the agent supports syncing a Git repository in the same cluster where it is deployed.

The agent has two modes we can use:

- **Namespaced**: The agent only manages the resources in the same namespace where it is deployed.
- **Full cluster**: The agent manages the whole cluster where it is deployed.

First, let's install the agent in a namespaced mode using the following commands:

```
kubectl create ns example-namespaced
kubectl apply -f https://raw.githubusercontent.com/
PacktPublishing/ArgoCD-in-Practice/main/ch09/namespaced/
namespaced.yaml

serviceaccount/gitops-agent created
role.rbac.authorization.k8s.io/gitops-agent created
rolebinding.rbac.authorization.k8s.io/gitops-agent created
deployment.apps/gitops-agent created
```

Now, let's check the status of the deployment:

```
kubectl -n example-namespaced rollout status deploy/gitops-
agent

Waiting for deployment "gitops-agent" rollout to finish: 0 of 1
updated replicas are available...
deployment "gitops-agent" successfully rolled out
```

If we check the logs of the GitOps agent, we will see that the synchronization finished successfully:

```
"msg"="Synchronization triggered by API call"
"msg"="Syncing" "skipHooks"=false "started"=false
"msg"="Tasks (dry-run)" "tasks"=[{}]
"msg"="Applying resource Deployment/app in cluster:
"https://10.96.0.1:443, namespace: example-namespaced"
"msg"="Updating operation state. phase:  -> Running, message:
'' -> 'one or more tasks are running'"
"msg"="Applying resource Deployment/app in cluster:
https://10.96.0.1:443, namespace: example-namespaced"
"msg"="Adding resource result, status: 'Synced', phase:
'Running', message: 'deployment.apps/app created'"
"kind"="Deployment" "name"="app" "namespace"="example-
namespaced" "phase"="Sync"
```

```
"msg"="Updating operation state. phase: Running -> Succeeded,
message: 'one or more tasks are running' -> 'successfully
synced (all tasks run)'"
```

We can see that there are two new Pods based on the example deployment we pointed the GitOps agent toward:

```
NAME                             READY    STATUS     RESTARTS    AGE
app-f6c66b898-gxcmn              1/1      Running    0           4s
app-f6c66b898-ljxdq              1/1      Running    0           4s
gitops-agent-648cf56fc8-2fxpr    2/2      Running    0           12s
```

Figure 9.1 – The GitOps agent namespaced mode successfully completed

If we check the YAML manifest we used, we will see that we have deployed two containers. The first container is about the GitOps agent and the second is k8s git-sync, which pulls a Git repository in a local directory, and we mount this directory to the GitOps agent container. Specifically, it's defined as the following:

```
- image: argoproj/gitops-agent:latest
  name: gitops-agent
  command:
  - gitops
  - /tmp/git/repo
  - --path
  - ch09/namespaced/example
  - --namespaced
  volumeMounts:
  - mountPath: /tmp/git
    name: git
```

The preceding is the container of the GitOps agent, which will use the ch09/namespaced/example path to sync and deploy the relevant Kubernetes resources included in this directory:

```
- image: k8s.gcr.io/git-sync:v3.1.6
  name: git-sync
  args:
  - --webhook-url
  - http://localhost:9001/api/v1/sync
  - --dest
```

```
  - repo
  - --branch
  - main
  env:
  - name: GIT_SYNC_REPO
    value: https://github.com/PacktPublishing/ArgoCD-in-
Practice
  volumeMounts:
  - mountPath: /tmp/git
    name: git
```

The preceding is a sidecar container of the Kubernetes `git-sync`, which will use the Git repository at `https://github.com/PacktPublishing/ArgoCD-in-Practice` to sync and deploy the relevant Kubernetes resources included in this directory.

Similarly, we can deploy the GitOps agent in full cluster mode by removing the `--namespaced` CLI flag, and then the GitOps agent can manage Kubernetes resources in any cluster namespace.

We have a full picture of what GitOps Engine is and its components and we got our first view of the GitOps agent and how we can use it for very simple cases. It's time to wrap up what we learned in this book and what the future is after this.

What is kubernetes-sigs/cli-utils?

In parallel with the Argo team, a Kubernetes SIG has been formed to cover topics such as `kubectl` and focus on the standardization of the CLI framework and its dependencies. This SIG group created a set of Go libraries called `cli-utils` in order to create an abstraction layer for `kubectl` and it has evolved to support server-side use in GitOps controllers too. The core features of `cli-utils` for GitOps are the following:

- Pruning
- Status interpretation
- Status lookup
- Diff and preview
- Waiting for reconciliation
- Resource ordering
- Explicit dependency ordering
- Implicit dependency ordering
- Applying time mutation

kapply usage

The Kubernetes SIG team for `cli-utils` created a CLI called `kapply`, which is not intended to be for production use, but it gives us a chance to understand the features and how we can better utilize the set of libraries provided in it.

Let's now see an example of using `kapply` and `cli-utils` by deploying a simple service with them. First, we need to clone the `cli-utils` repo and build the `kapply` CLI with the following command:

```
git clone git@github.com:kubernetes-sigs/cli-utils.git
make build
```

Validate that the CLI is working with the following command:

```
kapply --help
Perform cluster operations using declarative configuration

Usage:
  kapply [command]
Available Commands:
  apply        Apply a configuration to a resource by package
directory or stdin
  completion   Generate the autocompletion script for the
specified shell
  destroy      Destroy all the resources related to
configuration
  diff         Diff local config against cluster applied version
  help         Help about any command
  init         Create a prune manifest ConfigMap as a inventory
object
  preview      Preview the apply of a configuration
  status
```

We will create an example Kubernetes deployment and use the `Preview` feature to see which commands will be executed. You can find the example deployment in the Git repo under the `ch09/cli-utils-example/deployment.yaml` path. First, we need to create the inventory template that will be used in order to track the namespace and the inventory ID for the inventory objects in the Kubernetes deployment:

```
cd ch09/cli-utils-example
kapply init $PWD
namespace: example-cli-utils is used for inventory object
```

This command will create a file called `inventory-template.yaml` and its Kubernetes `ConfigMap`, which will observe the namespace we see in the output and the inventory ID. Now, let's run the `Preview` feature command and observe the output:

```
kapply preview $PWD

Preview strategy: client
inventory update started
inventory update finished
apply phase started
deployment.apps/app apply successful
apply phase finished
inventory update started
inventory update finished
apply result: 1 attempted, 1 successful, 0 skipped, 0 failed
```

The output gives us a preview of the commands that are going to run, but nothing is applied yet and this runs only on the client side. So, if we try to see all the resources in the Kubernetes cluster, we should see no resources created. The next step is to run the same command with server validation, which means a preview of the Kubernetes cluster:

```
kapply preview $PWD --server-side

Preview strategy: server
inventory update started
inventory update finished
apply phase started
deployment.apps/app apply failed: namespaces "example-cli-
utils" not found
apply phase finished
inventory update started
inventory update finished
apply result: 1 attempted, 0 successful, 0 skipped, 1 failed
E0929 00:49:04.607170    47672 run.go:74] "command failed"
err="1 resources failed"
```

In this case, we see that validation with the Kubernetes server returns an error. The reason is that the namespace to which the deployment manifest refers doesn't exist. Let's create the namespace and then try to apply the deployment:

```
kubectl create namespace example-cli-utils
namespace/example-cli-utils created
# Apply the deployment
kapply apply $PWD --reconcile-timeout=1m --status-events

inventory update started
inventory update finished
apply phase started
deployment.apps/app apply successful
apply phase finished
reconcile phase started
deployment.apps/app reconcile pending
deployment.apps/app is InProgress: Replicas: 0/2
deployment.apps/app is InProgress: Replicas: 0/2
deployment.apps/app is InProgress: Replicas: 0/2
deployment.apps/app is InProgress: Available: 0/2
deployment.apps/app is InProgress: Available: 1/2
deployment.apps/app is Current: Deployment is available.
Replicas: 2
deployment.apps/app reconcile successful
reconcile phase finished
inventory update started
inventory update finished
apply result: 1 attempted, 1 successful, 0 skipped, 0 failed
reconcile result: 1 attempted, 1 successful, 0 skipped, 0
failed, 0 timed out
```

In the output, we see all the phases have been passed to the successful deployment in the cluster. Now, let's change the number of replicas in the deployment manifest from 2 to 3, apply it again, and see the output:

```
kapply apply $PWD --reconcile-timeout=1m --status-events

deployment.apps/app is Current: Deployment is available.
Replicas: 2
```

```
inventory update started
inventory update finished
apply phase started
deployment.apps/app apply successful
apply phase finished
reconcile phase started
deployment.apps/app reconcile pending
deployment.apps/app is InProgress: Deployment generation is 2,
but latest observed generation is 1
deployment.apps/app is InProgress: Replicas: 2/3
deployment.apps/app is InProgress: Replicas: 2/3
deployment.apps/app is InProgress: Available: 2/3
deployment.apps/app is Current: Deployment is available.
Replicas: 3
deployment.apps/app reconcile successful
reconcile phase finished
inventory update started
inventory update finished
apply result: 1 attempted, 1 successful, 0 skipped, 0 failed
reconcile result: 1 attempted, 1 successful, 0 skipped, 0
failed, 0 timed out
```

In the output, we clearly observe that there is a difference between the manifest state and the cluster state and this is the reason that the status is reconciled as pending.

This was a demonstration of one of the multiple features provided by `cli-utils` and gave us a good view of the capabilities and how we can leverage them in our own implementation and use cases. In the next section, we will summarize what we have learned in this book and what is next after this book.

Wrap up

When Argo CD started back in 2018, nobody could have predicted its success. It had a good foundation (the Application CRD with its source, the Git repository where the manifests are located, and the destination, which is the cluster and namespace where deployments are performed), was well received, understood, and a good fit for the whole GitOps concept.

It also had the right context. Back then, as now, Helm was the most used application deployment tool for Kubernetes, and it was in the V2 version. This meant it came with a component called Tiller installed on the cluster (`https://helm.sh/docs/faq/changes_since_helm2/#removal-of-tiller`), which was used to apply the manifests, and that component was seen as a big security hole. With Argo CD, you could have still used Helm charts, but you didn't need Tiller to perform the installation as manifests were generated and applied to the destination cluster by a central Argo CD installation. I remember back then we saw this as a major benefit.

There were many other features that put it in front of the competition, such as a great UI where you could see how deployments were happening in your cluster (which also became a good tool for Kubernetes onboarding), a good CLI (which operators like myself were able to use for troubleshooting, providing more info than the UI), or single sign-on, (which opened the path to adoption in the enterprise world, as having each tool with its own access control was considered too big of a risk to have it wrongly configured).

Fast forward 4 years, and these days (October 2022), the project has almost 11,000 stars and over 800 contributors, showing a healthy ecosystem that most likely is going to grow in the next few years. The Argo umbrella includes four main projects: Argo CD, Argo Workflows, Argo Events, and Argo Rollouts, and they are now part of the incubating projects in CNCF. These are considered mature and stable ones, and with relevant usage in production. There is only one more step for Argo, and that is to reach the graduated stage, to join projects such as Kubernetes, Prometheus, Helm, or Envoy, which would mean an even wider adoption and more contributors – things that will probably happen in the next few years.

We already had the second edition of ArgoCon (`https://www.youtube.com/playlist?list=PLj6h78yzYM2MbKazKesjAx4jq56pnz1XE`) in September 2022, which was the first one with on-site attendance, as 2021 was only online. ArgoCon is a vendor-neutral event, organized by the community and following CNCF best practices, such as having a proper program committee. We also have two companies now offering a commercial Argo CD platform, which is again a good sign of a healthy ecosystem. We can mention that this book is a step in the same direction.

All these show that now is a good time to consider adopting GitOps and not hesitate to give Argo CD a chance.

Summary

Congrats on reading and participating in the practical parts of the book – it was a really long road and now you are reading the last section of the book.

In this book, we started by helping you get familiarized with GitOps, Kubernetes, and Kubernetes operators. After this, we continued with more advanced topics and real-life implementations for introducing Argo CD and its components, bootstrapping a Kubernetes cluster in a repeatable manner, operating and troubleshooting Argo CD, making Argo CD easy to be adapted in enterprise organizations with access control features, and designing production-ready Argo CD delivery pipelines with complex deployment strategies.

In each chapter, we presented you the code and the inspiration to try building and implementing your own solutions on Argo CD and how it can help us to implement complex orchestrate scenarios with Sync Waves and phases and enable complex deployment strategies with Argo Rollouts.

Finally, we went through the democratization of GitOps and how GitOps Engine or `kubernetes-sigs/cli-utils` can set the basis for the future of OpenGitOps and the standardized approach to implementing GitOps.

The Argo project is a suite of services and in the future, it may turn out to be a good idea to develop a book to present an integration between them that uses Argo CD, Argo Rollouts, Argo Events, and Argo Workflows, leveraging the full suite in an end-to-end solution.

Further reading

- Democratizing GitOps – `https://blog.argoproj.io/argo-cd-v1-6-democratizing-gitops-with-gitops-engine-5a17cfc87d62`

- GitOps Engine – `https://github.com/argoproj/gitops-engine`

- The GitOps agent – `https://github.com/argoproj/gitops-engine/tree/master/agent`

- `cli-utils` – `https://github.com/kubernetes-sigs/cli-utils`

Index

A

`Packt.com`

Subscribe to our online digital library for full access to over 7,000 books and videos, as well as industry leading tools to help you plan your personal development and advance your career. For more information, please visit our website.

Why subscribe?

- Spend less time learning and more time coding with practical eBooks and Videos from over 4,000 industry professionals

- Improve your learning with Skill Plans built especially for you

- Get a free eBook or video every month

- Fully searchable for easy access to vital information

- Copy and paste, print, and bookmark content

Did you know that Packt offers eBook versions of every book published, with PDF and ePub files available? You can upgrade to the eBook version at `packt.com` and as a print book customer, you are entitled to a discount on the eBook copy. Get in touch with us at `customercare@packtpub.com` for more details.

At `www.packt.com`, you can also read a collection of free technical articles, sign up for a range of free newsletters, and receive exclusive discounts and offers on Packt books and eBooks.

Other Books You May Enjoy

If you enjoyed this book, you may be interested in these other books by Packt:

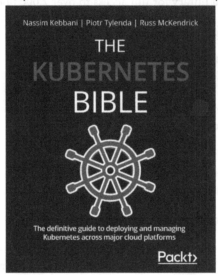

The Kubernetes Bible

Nassim Kebbani, Piotr Tylenda, Russ McKendrick

ISBN: 978-1-83882-769-4

- Manage containerized applications with Kubernetes
- Understand Kubernetes architecture and the responsibilities of each component
- Set up Kubernetes on Amazon Elastic Kubernetes Service, Google Kubernetes Engine, and Microsoft Azure Kubernetes Service
- Deploy cloud applications such as Prometheus and Elasticsearch using Helm charts
- Discover advanced techniques for Pod scheduling and auto-scaling the cluster
- Understand possible approaches to traffic routing in Kubernetes

Learning DevOps - Second Edition

Mikael Krief

ISBN: 978-1-80181-896-4

- Understand the basics of infrastructure as code patterns and practices
- Get an overview of Git command and Git flow
- Install and write Packer, Terraform, and Ansible code for provisioning and configuring cloud infrastructure based on Azure examples
- Use Vagrant to create a local development environment
- Containerize applications with Docker and Kubernetes
- Apply DevSecOps for testing compliance and securing DevOps infrastructure
- Build DevOps CI/CD pipelines with Jenkins, Azure Pipelines, and GitLab CI
- Explore blue-green deployment and DevOps practices for open sources projects

Packt is searching for authors like you

If you're interested in becoming an author for Packt, please visit `authors.packtpub.com` and apply today. We have worked with thousands of developers and tech professionals, just like you, to help them share their insight with the global tech community. You can make a general application, apply for a specific hot topic that we are recruiting an author for, or submit your own idea.

Share Your Thoughts

Now you've finished *Argo CD in Practice*, we'd love to hear your thoughts! Scan the QR code below to go straight to the Amazon review page for this book and share your feedback or leave a review on the site that you purchased it from.

https://packt.link/r/180323332X

Your review is important to us and the tech community and will help us make sure we're delivering excellent quality content.

Download a free PDF copy of this book

Thanks for purchasing this book!

Do you like to read on the go but are unable to carry your print books everywhere?

Is your eBook purchase not compatible with the device of your choice?

Don't worry, now with every Packt book you get a DRM-free PDF version of that book at no cost.

Read anywhere, any place, on any device. Search, copy, and paste code from your favorite technical books directly into your application.

The perks don't stop there, you can get exclusive access to discounts, newsletters, and great free content in your inbox daily

Follow these simple steps to get the benefits:

1. Scan the QR code or visit the link below

https://packt.link/free-ebook/9781803233321

2. Submit your proof of purchase
3. That's it! We'll send your free PDF and other benefits to your email directly

www.ingramcontent.com/pod-product-compliance
Lightning Source LLC
Chambersburg PA
CBHW060547060326
40690CB00017B/3635